The Family Tree Guide to DNA Testing and Genetic Genealogy

The Family Tree Guide to DNA Testing and Genetic Genealogy

Blaine T. Bettinger

CINCINNATI, OHIO
shopfamilytree.com

CONTENTS

PART THREE: ANALYZING AND APPLYING TEST RESULTS

INTRODUCTION

Genealogical records are not perfect. Our ancestors had poor memories just like we do: They bent the truth to make themselves younger or seem more favorable just like we do, and they made up stories just like we do. In addition, genealogical records can be altered, improperly transcribed or recorded, or completely lost, even in the years immediately after their creation.

As a result, genealogists work with a trail of imperfect and inconsistent bread crumbs, and we use these traces to recreate the lives of our ancestors. Sometimes we do a good job, sometimes we do a poor job, and sometimes we may not know the difference.

Trapped within your DNA, however, are the stories of your ancestors. Although this information was inaccessible to previous generations of genealogists, modern genetic testing has allowed us to extract those stories and begin to add them to the rich genealogical tapestry so many of us spend our lives creating.

Although DNA is a (mostly) unchangeable record of those ancestors who provided bits and pieces of their DNA to the current generation, we are currently limited in our ability to properly interpret the whole record. Indeed, interpreting DNA test results can introduce errors and inconsistencies. As a result, current genetic genealogy testing is not yet a perfect genealogical record either. Only when DNA and traditional genealogical records are combined do we begin to fully extract the full value of genetic testing.

My Genetic Genealogy Journey

I was introduced to genealogy in the seventh grade. My English teacher assigned a short homework assignment: a four- or five-generation family tree that we were to fill out as far back as we could by asking family members for information. My parents recommended that I call my grandmother, as she was one of the oldest members of that generation. During the call, my grandmother recited a litany of names and places completely from memory. I frantically filled out the paternal side of the family tree, then added sheets of paper with even more ancestors. With that one phone call, I was hooked. I've spent the ensuing twenty-five years trying to verify those names, learn about those ancestors' lives, and fill in the blanks that my grandmother couldn't provide.

I took my first DNA test in 2003. I was a biochemistry graduate student in Syracuse, New York, and a DNA test was the perfect marriage of the two great loves of my life: genealogy and science. Unlike most other people in that timeframe, who started their DNA journey with a Y-chromosomal (Y-DNA) or mitochondrial-DNA (mtDNA) test, my first test was an autosomal DNA (atDNA) test. It only looked at a little over one hundred markers (compared to the hundreds of thousands of markers used today), but I was once again hooked. That autosomal test was just the first in a very long line of DNA testing, including a full sequence of my entire genome by the Personal Genome Project ten years later.

Along the way, I've learned a great deal about my genealogical heritage as a result of DNA testing. I know that my mitochondrial DNA is Native American, meaning that my mother's mother's mother's mother was Native American at some point. I know that I carry African and Native American segments of DNA from Central American ancestors. I know that I have pieces of DNA passed down to me from my French Canadian and Irish ancestors. And I am beginning to identify the parents of my adopted great-grandmother,

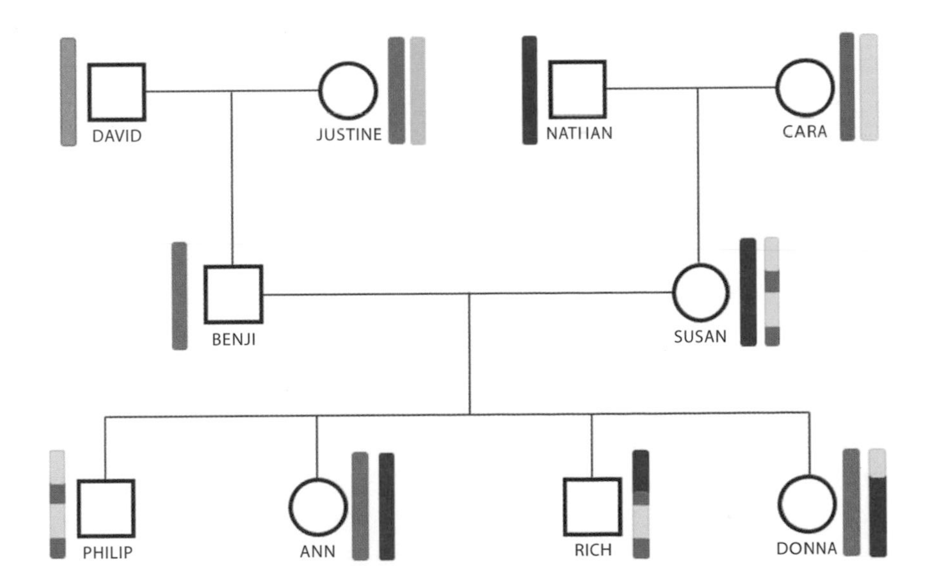

Throughout the book, diagrams, such as this X-DNA chart from chapter 7, will help illustrate important concepts in genetic genealogy.

all due to DNA testing. I've used modern genetic testing to extract stories that I was carrying with me all along—stories that my ancestors, including the grandmother who gave me the gift of genealogy, unknowingly passed down to me.

Across the country and across the world, every vial of spit and every cheek swab thrown in the mail with anticipation is overflowing with long-forgotten stories waiting to be unlocked from within a series of *A*'s, *T*'s, *C*'s, and *G*'s.

How to Use This Book

This book is intended to be a resource for genealogists of all experience levels, from beginner to expert and everyone in between. If you've never taken a DNA test, then you can use this book as a primer to understand what DNA testing is, whether DNA testing is right for you, and how you can use the results of DNA testing to examine your ancestry. My advice to you is to read the book from start to finish as it is presented, since it is written with more basic information at the beginning and more advanced information toward the end.

Throughout the book, you'll note special terms, indicated in red. These are important vocabulary words you'll encounter in your genetic research, and I've compiled them into a glossary toward the back of the book for your reference.

If you've already taken a DNA test, you can use this book as a reference while you review your test results. After making a quick stop at chapters 1 (basics), 2 (common misconceptions), and 3 (ethics), turn to the chapter relevant to your type of DNA testing: mtDNA (chapter 4), Y-DNA (chapter 5), or atDNA (chapters 6-9). Then, read the remaining chapters to ensure you've filled in all the blanks and have a complete understanding of the different aspects of genetic genealogy testing.

A genealogist's education is never complete. It is important that genealogists stay on top of the latest developments in DNA testing and analysis. Accordingly, if you feel you've mastered most of the topics presented in this book, turn to the More Resources section in Appendix C for links to some of the best blogs, forums, and mailing lists available. These links will allow you to discover and explore all the latest developments and advances in genetic genealogy.

I wish you the best of luck as you embark on your own DNA journey!

Blaine T. Bettinger

April 2016

PART ONE

Getting Started

1 Genetic Genealogy Basics

Genealogists are family historians, documenting known information about a family and using historical records to recreate and recover information that has been lost due to time and distance. As a result, personal possessions like family Bibles, wartime letters, and dusty daguerreotypes are among the most treasured items a genealogist can receive from an ancestor or relative. These possessions are often unique and reveal information that may otherwise be lost. Passing down these treasured family records and keepsakes is an important tradition that preserves memories for future generations.

However, prior generations have been passing down more than memories and keepsakes to their descendants. At every generation, our ancestors passed down indelible records through their DNA, pieces of themselves that they received from their own ancestors. That inheritance is the reason you have your great-aunt Millie's curly hair, your grandfather's heavy eyebrows, or your great-grandmother's deep blue eyes.

You are the keeper of your ancestors' DNA, and with a new tool called **genetic genealogy**, you can unlock that DNA and reveal the secrets it has safeguarded for generations. Indeed, even adoptees who have no knowledge of their biological ancestors can use this tool to find genetic relatives and learn about their biological heritage.

A

Genetic genealogy was first used to examine historical and forensic questions, such as if certain remains belonged to the Tsarina Alexandra.

In this chapter, we'll outline the history and basics of genetic genealogy to prep you for the rest of the book. As you read, you'll learn more about the different types of genetic genealogy testing and how you can use your results to examine your heritage, answer genealogical questions, and solve family mysteries. You'll learn about which test(s) you should take and what limitations you should keep in mind when reviewing your test results. You'll also discover (among many other things) some of the third-party tools you can use to wring every bit of useful information from your DNA test.

The History of Genetic Genealogy

Before genealogists used it, genetic genealogy was utilized by scientists and historians to identify genealogical connections between high-profile historical figures. In 1994, for example, mitochondrial-DNA (mtDNA) testing—one of the first tests to become available—was used to identify skeletons found in 1991 in a shallow grave in Ekaterinburg,

CHARLES DARWIN AND GENETIC GENEALOGY

In the mid-1800s, Charles Darwin first proposed the groundbreaking theories of evolution and natural selection, upon which much of modern genetics is based. Nearly two hundred years later, Darwin's own genetic roots were examined with a simple DNA test. In early 2010, National Geographic's Genographic Project tested the Y-DNA of Darwin's great-great-grandson Chris Darwin of Australia. The test revealed that Chris, and thus most likely Charles, belong to the *R1b* haplogroup, the most common haplogroup in males of European descent. (We'll discuss haplogroups in more detail later.)

Russia, as those of the Romanov family who were killed in 1918. Using a low-resolution test, scientists discovered that mtDNA extracted from several of the skeletons (including those hypothesized to be the Tsarina Alexandra in image A—a maternal granddaughter of Queen Victoria—and several of the Tsarina's children) matched mtDNA obtained from Prince Philip, Duke of Edinburgh, who is a great-great-grandchild of Queen Victoria. Since that early testing, both mtDNA testing and autosomal-DNA (atDNA) testing (another popular type of testing) have identified the remains of Tsar Nicholas II and his entire family, including the Tsarina and all five of their children.

In a similar way, Y-chromosomal (Y-DNA) testing was used in 1998 to show a genetic match between a male relative of President Thomas Jefferson and a descendant of Eston Hemings, the youngest son of Thomas Jefferson's slave Sally Hemings. The descendants of Eston Hemings had a strong oral tradition that Eston's father was indeed Thomas Jefferson, and Eston was said to bear a strong resemblance to Jefferson. Many historians, however, believed that Eston's father was one of the sons of Jefferson's sister, which might explain the resemblance. Since the former president had no surviving legitimate sons to pass down Y-DNA to, researchers obtained a Y-DNA sample from five male-line descendants of Jefferson's paternal uncle, Field Jefferson. Those samples were compared to Y-DNA samples obtained from a living descendant of Eston Hemings, and the two were a genetic match. Today, many historians accept that Jefferson fathered several children with Sally Hemings—including Eston.

Recognizing the power of DNA to examine genealogical relationships, genealogists began to investigate ways to use the tool. A few years after historians successfully used DNA testing to reveal Jefferson's descendants, a group of scientists including a man named Bryan Sykes conducted a study examining the Y-DNA of forty-eight males in the United Kingdom with the last name *Sykes*. The low-resolution Y-DNA testing determined that almost half of the males were related through their paternal or (surname) line, suggesting a single surname founder for these males. The scientists noted that Y-DNA studies such as the one they had conducted could have numerous applications in forensics and genealogy.

Eventually, the practical implications of DNA testing for genealogists became clear. In early 2000, two companies began offering DNA testing to genealogists: Family Tree DNA **<www.familytreedna.com>**, based in Houston, Texas, and led by Bennett Greenspan, Max Blankfeld, and Jim Warren; and Oxford Ancestors **<www.oxfordancestors.com>**, based in Oxfordshire, England, and created by Bryan Sykes of the Sykes surname study. Both companies launched by offering Y-DNA and mtDNA testing to genealogists, the first such commercial products.

Over the next few years, genetic genealogy testing expanded widely, led by large projects that combined Y-DNA testing and surnames, similar to the Sykes study's methodology. In the fall of 2007, genetic genealogy testing company 23andMe **<www.23andme.com>** began offering the first commercial atDNA test, and in 2012, AncestryDNA **<www.dna.ancestry.com>** officially launched its own atDNA test. Today, 23andMe, AncestryDNA, and Family Tree DNA still offer genetic testing to genealogists of all experience levels and are the leading genetic-genealogy companies. We'll learn about these and other testing companies in a later chapter.

Genetic Genealogy Today

Genetic genealogy is an essential tool for genealogists. It is an important piece of evidence similar to a census record, will, or land record, and it might be the last piece of information available in locations where records have been lost or destroyed. Although DNA testing cannot answer (or even shed light) on every question, savvy genealogists should at least consider it as part of every genealogical research project.

In the summer of 2015, 23andMe and AncestryDNA each announced that they had tested their one millionth customer, and their customer base is growing with thousands of new tests being sold each month. Although the Family Tree DNA database has traditionally been smaller than 23andMe's and AncestryDNA's databases, it is undeniably large and continues to grow rapidly.

As the databases' sizes grow, so does the power of genetic genealogy. New connections, tools, and discoveries will be made possible as more and more people take DNA tests.

A Little Genetics: What is DNA?

You don't need to have an advanced degree in molecular biology or genetics to understand genetic genealogy. You don't even need to remember anything from that biology course you took in the tenth grade. This brief introduction—and some details provided in each individual chapter—will be more than enough to help you understand how to use genetic genealogy testing for your research project.

The cell, the basic unit of life, uses genetic material called DNA to control the vast majority of its functions, beginning with the division of its parent cells and ending with its ultimate death. **DNA** (short for deoxyribonucleic acid) is a component of the cell that carries the instructions for the development and operation of all living things. A small percentage of the DNA comprises **genes**, short segments of DNA that are used as the blueprints to create a protein or an RNA (ribonucleic acid) molecule. Scientists also continue to find secondary functions for the **non-coding regions** of DNA, which don't specifically create proteins or RNA.

A molecule of DNA is composed of a string of millions of smaller units called **nucleotides**. Together, two intertwined DNA molecules interact to form a single double-helix structure called a **chromosome** in the **nucleus**—or control center—of the cell.

A normal human cell has ninety-two long molecules of DNA that pair up to form forty-six double-stranded chromosomes. Each of these, in turn, forms a **chromosome pair** with another similar—but not identical—chromosome, to create twenty-three different chromosome pairs.

Confused? Here's a table that breaks down the different levels of organization of DNA:

Component	Description
Nucleotide	The building block of DNA, it comes in four types that pair up in specific ways: adenine, cytosine, guanine, and thymine
DNA (deoxyribonucleic acid)	A double-stranded molecule comprising two entwined strings of millions of different nucleotides
Gene	A region of DNA along a chromosome that encodes for a functional product such as a protein
Chromosome	A highly organized double helix of two DNA molecules
Chromosome pair	Two complementary chromosomes, one inherited from each parent

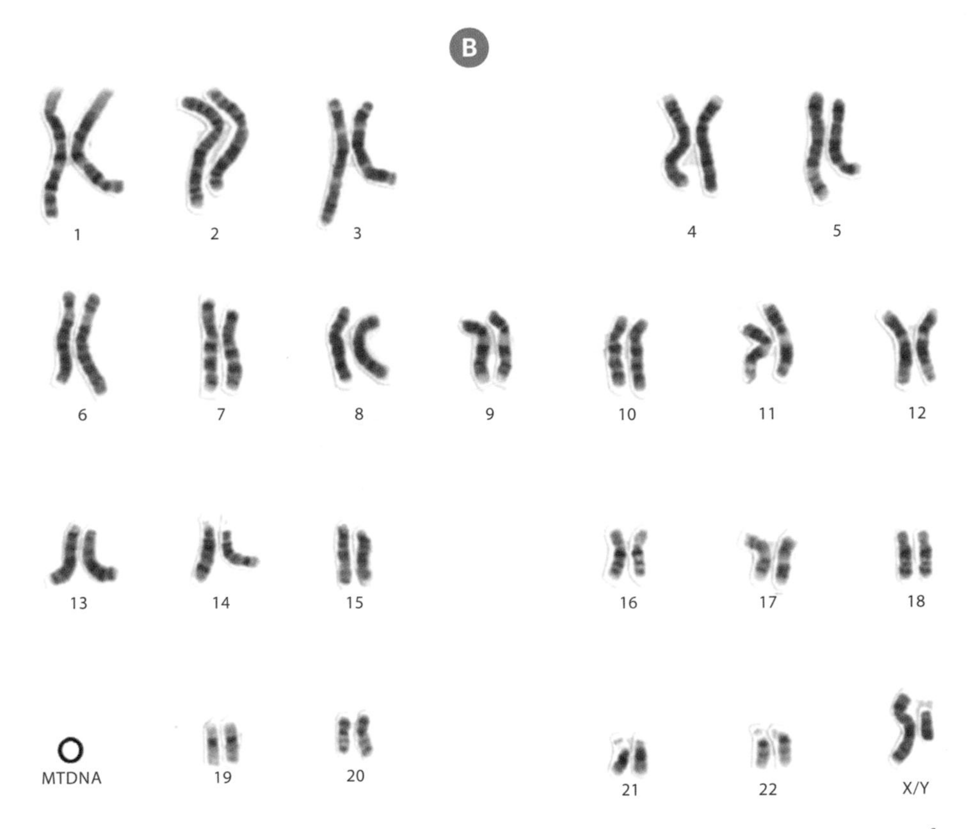

Each person has a unique genetic makeup, comprising twenty-two sets of chromosomes, a pair of sex chromosomes, and rings of mtDNA. These come together to form the human karyotype. This photo is courtesy of Darryl Leia of the National Human Genome Research Institute.

In addition to the DNA in the nucleus, hundreds or thousands of copies of a very small circular strand of DNA are found in the many mitochondria outside the nucleus. **Mitochondria** are tiny powerhouses of the cell responsible for, among other things, creating the energy our cells need to function.

Image B, a karyogram, is a photograph of a human's **karyotype**, which is all of the chromosomes of the human cell arranged in pairs in a numbered sequence from longest to shortest. To make a karyogram, researchers stain chromosomes with a special chemical, then take a photograph of the stained chromosomes. The chromosomes are then digitally rearranged into pairs and organized into a specific numbered sequence. This karyogram also includes a ring of mtDNA for reference.

In this book, you will examine the four types of DNA used for genetic genealogy: mtDNA, Y-DNA, atDNA, and X-DNA.

1. **Mitochondrial DNA (mtDNA)** is a small, circular piece of DNA found in the cell's energy factory, the mitochondria. This is the only DNA not found in the cell's nucleus. mtDNA is passed exclusively from mother to child, and an mtDNA test reveals information about the test-taker's direct maternal (or "umbilical") line. Chapter 4 focuses on mtDNA testing.
2. **Y-chromosomal DNA (Y-DNA)** focuses on the Y chromosome, one of the two sex chromosomes that determine gender (the other being the X chromosome). Only men have a Y chromosome, and a Y-DNA test reveals information about the (male) test-taker's Y chromosome, which is exclusively passed from fathers to sons. We'll go into more detail about the Y-DNA test in chapter 5.
3. **Autosomal DNA (atDNA)** is composed of pairs of chromosomes found in the nucleus of the cell. Humans have twenty-three pairs of chromosomes (forty-six total), of which twenty-two are autosomal DNA (or "autosomes") and one is sex chromosomes. One copy of each chromosome is inherited from the mother and one copy from the father. An atDNA test reveals information about both paternal and maternal lines, and we'll discuss this test more in chapter 6.
4. **X-chromosomal DNA (X-DNA)** focuses on the X chromosome, one of the two sex chromosomes that determine gender (the other being the Y chromosome). Women have two X chromosomes, one from their father and one from their mother; men have one X chromosome from their mother. X-DNA is usually tested as part of an atDNA test. For men, the X-DNA test (the subject of chapter 7) reveals information about maternal lines. For women, the X-DNA test reveals information about both maternal and paternal lines.

Two Family Trees: One Genealogical and One Genetic

One of the most important aspects of understanding and interpreting DNA test results is that everyone has two very different (but overlapping) family trees: one that's genealogical (reflecting familial relationships) and one that's genetic (reflecting genetic makeup and patterns of inheritance). In short, your genealogical family tree will contain everyone in your genetic family, but not vice versa.

The Genealogical Family Tree

The first—and probably best-known and most-studied—family tree is the **genealogical family tree**, which contains every ancestor who had a child who had a child who had a child, and so on. A fully grown genealogical tree (image C) contains every parent, grand-

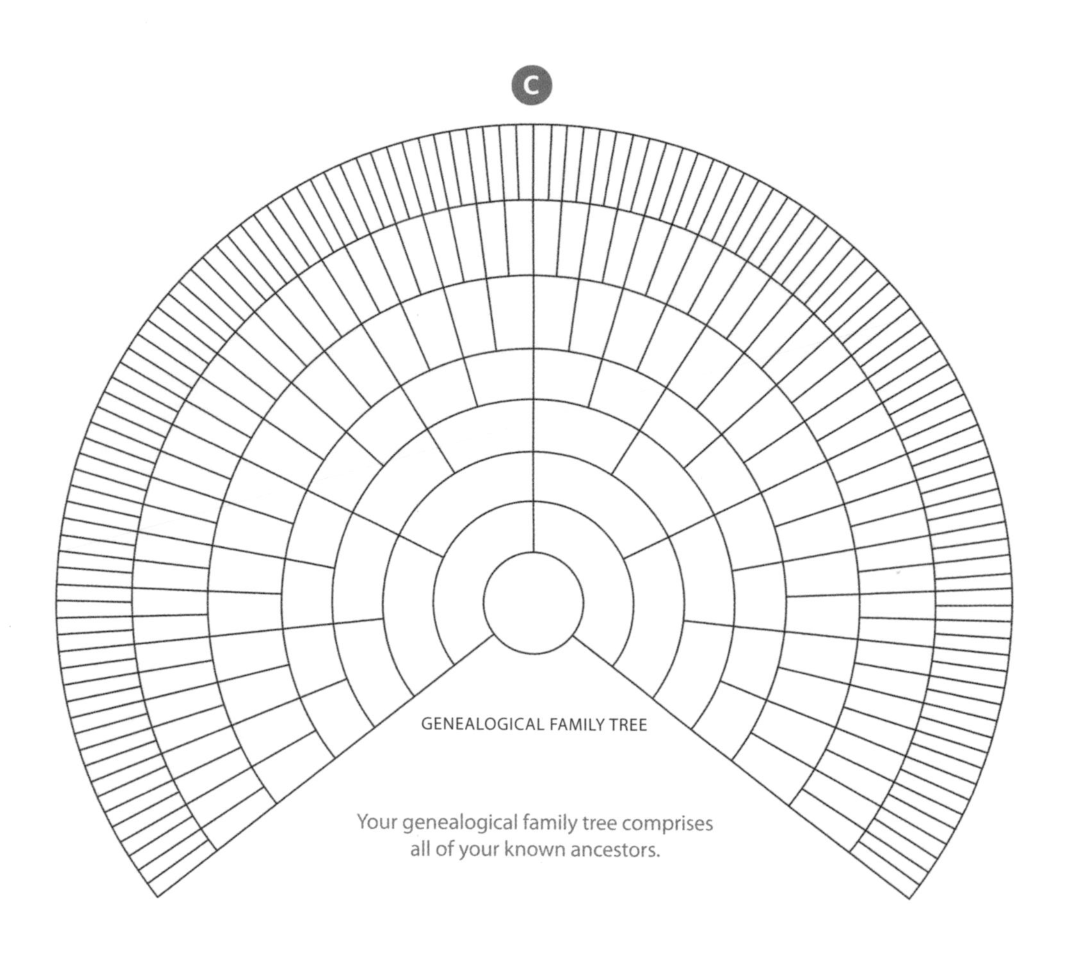

Your genealogical family tree comprises all of your known ancestors.

parent, and great-grandparent back through history. In most cases, this is the tree that genealogists spend their time researching, often using paper records such as birth and death certificates, census records, and newspapers to fill it in. Many genealogists find that the paper trail ends or becomes much more difficult to identify beyond the 1800s or 1700s, making it difficult to fill in many of the openings in the genealogical family tree.

The Genetic Family Tree

The second family tree is the **genetic family tree**, which contains only those ancestors who contributed to your DNA. While this overlaps with your genealogical family tree, not every person in a genealogical family tree contributes a segment of his or her DNA sequence to the test-taker's DNA sequence. A parent does not pass on all his DNA to his children (only about 50 percent); as a result, bits and pieces of DNA are lost in each generation. Your genetic family tree likely contains fewer ancestors than your genealogical family tree somewhere between five and nine generations back.

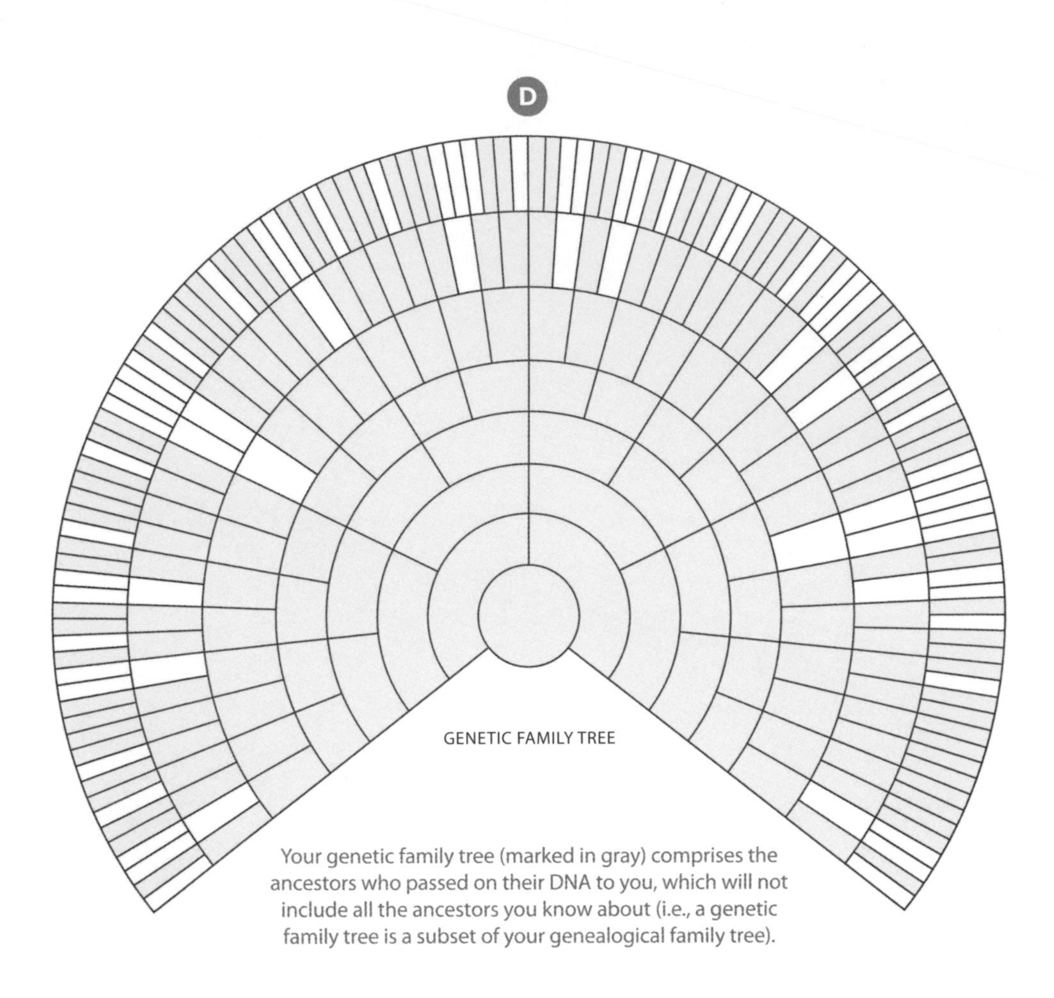

Your genetic family tree (marked in gray) comprises the ancestors who passed on their DNA to you, which will not include all the ancestors you know about (i.e., a genetic family tree is a subset of your genealogical family tree).

As shown in image D, where gray cells indicate that the ancestor provided DNA to the test-taker and white cells indicate that the ancestor did not provide DNA to the test-taker, the genetic tree is actually just a subset of your genealogical tree. A genetic tree is guaranteed to contain both biological parents, who each contributed approximately 50 percent of the test-taker's entire DNA sequence. The genetic tree also likely contains each of the test-taker's four biological grandparents and eight biological great-grandparents, but it is much less likely with each generation that every person in the genealogical family tree contributed a piece of their DNA to the test-taker's DNA.

Since test-takers have a genetic family tree that is a subset of their genealogical family tree, a person will often share his genealogical family tree with another individual but their genetic family trees do not overlap. This simply means they did not both inherit the same DNA from their shared ancestor. These individuals are genealogical cousins but not genetic cousins, as genetic cousins share both a genealogical and a genetic link to one another (i.e., they share one or more recent individuals in their genetic family trees and therefore share detectable amounts of DNA from this common ancestor). Biological first cousins, for example, always share DNA and therefore will always be both genealogical cousins and genetic cousins.

Unfortunately, no one has yet been able to construct a very complete genetic family tree, due in part to the lack of extensive databases combining paper genealogy and genetics. With the recent development of tools that combine genetic testing with family trees, however, genetic genealogists are beginning to reconstruct portions of their genetic family trees.

CORE CONCEPTS: GENETIC GENEALOGY BASICS

- **Genetic genealogy started as a tool in history and forensic science. In the 1900s, DNA companies began offering genetic tests for use in genealogy.**

- **Genealogists use four different kinds of DNA in testing: mitochondrial DNA (mtDNA), Y-chromosomal DNA (Y-DNA), autosomal DNA (atDNA), and X-chromosomal DNA (X-DNA). You'll want to employ different tests depending on your research goals.**

- **Everyone has two sets of ancestors: a genealogical family tree (ancestral family members) and a genetic family tree (ancestors who contributed DNA). The genetic family tree is a subset of the genealogical family tree and can sometimes be hard to pinpoint.**

2

Common Misconceptions

DNA is a powerful tool for genealogists. It can confirm or correct family trees and pedigrees, find new relatives you never knew existed, and help you learn about your family's ancient origins. However, DNA is not magic. Just as a census record or deed alone cannot provide all the answers to your genealogical questions, DNA is not a miracle solution to all research problems. To be successful, genetic genealogists must diligently work with their DNA test results and combine them with other types of records to reach a defensible conclusion.

In many ways, scientists, prosecutors, and genealogists have oversold DNA as a cure-all solution for understanding health, solving crimes, and breaking through brick walls. As a result, people have many misconceptions about how DNA can be used. For example, even if law enforcement has an excellent genetic sample, DNA cannot solve all crimes or identify all suspects or perpetrators. Even if researchers spend billions of dollars on genetic research, DNA cannot explain all causes of disease or provide a cure for all illnesses. And even if genealogists have the best tools and enormous databases of test-takers to draw from, DNA cannot break through every brick wall. Understanding the limitations of DNA is vital during all aspects of DNA testing, including when creating a testing plan, reviewing results, drawing conclusions, and writing about or sharing results. Mastering these tips will help you avoid the mistakes that genetic genealogists most often make.

In this chapter, we'll address some of the most common misunderstandings about genetic genealogy and why they're incorrect. As we examine each individual type of DNA tests in later chapters, you will gain an even more in-depth understanding of the benefits and limitations of genetic genealogy.

MISCONCEPTION #1: Genetic genealogy is just for fun.

There is no doubt that genetic genealogy is a fun and interesting new way to explore genealogy. Popular mainstream television shows like *Finding Your Roots* and *Who Do You Think You Are?* use DNA testing to support and augment the powerful family stories shared by celebrity guests. Print and online advertisements from 23andMe **<www.23andme.com>** and AncestryDNA **<www.dna.ancestry.com>** popularize ethnicity estimates, leading many thousands of people to purchase DNA tests and explore their roots. Young people in particular have been drawn to genealogy in greater numbers than ever before due in no small part to genetic genealogy. But can genetic genealogy only be used for entertainment?

A genealogist should examine every possible record that can shed light on a genealogical question. If you wonder whether Great-great-grandfather Ned owned land in rural New York (where he lived), you should of course check land records. But you should also check tax records, probate records, and any other records that might be helpful. Accordingly, a genealogist should use DNA testing whenever it may shed light on a question or whenever it can support an existing conclusion or hypothesis. The Smith line may be the most well-documented line you've ever reviewed or constructed, but have you tested your conclusions with DNA? Are you certain there weren't any **non-paternal events**—breaks in the expected Y-DNA line due to adoption, infidelity, or other causes—that may not have shown up in the documentary evidence?

DNA testing is fun, but it's much more than a form of entertainment. It is a piece of evidence, just like any census record, vital record, tax record, probate record, or land record, that should be evaluated as a potential tool for every research question. Eventually, considering DNA testing should become as reflexive to genealogists as checking an ancestor's census returns and vital records.

MISCONCEPTION #2: I'm a woman, so I can't take a genetic genealogy test myself.

Contrary to popular belief, women can take three of the four major genetic tests, and both male and female genealogists can benefit from the results of all four.

This misconception is due to technical limitations in early genetic genealogy. Back in the days when genetic genealogy was primarily basic Y-DNA and mitochondrial-DNA (mtDNA) testing, genealogists were repeatedly told that only men can take a Y-DNA test. Although women have always been able to take an mtDNA test, the early form of this test was not as genealogically informative as a Y-DNA test. Indeed, most of the emphasis was on Y-DNA testing for the first ten years of genetic genealogy (2000–2010), and so some women in the genealogy community felt excluded.

But women can even participate in Y-DNA testing, albeit not directly. For example, a woman interested in her Y-DNA line could find another living (and willing) source of that DNA. Fathers, brothers, uncles, or male cousins are all potential sources of Y-DNA for testing. In some cases in which you can't find a father, uncle, or brother, the source may be several generations removed. But the secret to finding these sources is to do what every good genealogist does: Use documentary genealogical research to find a paternal line descendant who is willing to take a Y-DNA test.

Further, there is no limitation on who may take an autosomal-DNA (atDNA) test. atDNA examines many different lines of the family tree, rather than just the paternal (Y-DNA) and maternal (mtDNA) lines. Everyone has the same amount of atDNA and can take an atDNA test.

MISCONCEPTION #3:
DNA testing will provide me with a family tree.

One of the biggest misconceptions surrounding genetic genealogy is that the results of a DNA test are a magic bullet for revealing your family tree. Unfortunately, a DNA test alone does not provide a family tree (or, at least, none of the tests that are currently available). The test-taker does not log into her DNA testing account and see a partial or complete family tree as part of the results. Instead, as we'll see in chapter 6, the test-taker usually receives two categories of information: an ethnicity prediction and a list of genetic matches who share one or more segments of DNA with the test-taker.

DNA in conjunction with traditional research, however, is a powerful tool that can help you research and re-create your family tree. For example, the name of your great-grandmother is not directly encoded in your DNA, so simply analyzing your DNA with a test cannot reveal her name. However, clues to your great-grandmother's identity are encoded in your DNA; she gave you some of her DNA, and you share some of that DNA with your genetic and genealogical cousins. Through DNA testing, documentary research, and hard work, you can collaborate with these genetic cousins to identify your shared ancestry, which may include your great-grandmother or your great-grandmother's ancestor. This collaborative effort helps confirm branches of an existing family tree and helps break through brick walls.

Similarly, a genetic genealogy test can sometimes make it easy to find an existing family tree. Adoptees who take a DNA test and are able to connect with their biological families will also simultaneously receive biological family trees for one or both sides of their newly identified family. Although not all adoptees find their biological family after taking a DNA test, DNA testing increasingly is able to identify one or two parents of most adoptees.

To put it another way: Like most genealogical research, DNA testing cannot reach its full potential without context. In most cases, that context is the documentary research that the test-taker or the test-taker's genetic matches have performed on their family trees.

MISCONCEPTION #4: **DNA results are too narrow to be worthwhile.**

This misconception is most commonly found in news articles writing about genealogy testing. For example, articles often highlight the fact that several types of DNA testing can only reveal information about a tiny percentage of your ancestry. Indeed, a Y-DNA test only examines the direct male line (your father's father's father, and so on). Similarly, an mtDNA test only examines the direct female line (your mother's mother's mother, and so on). A ten-generation family tree contains up to 1,024 ancestors at the tenth generation, but a Y-DNA or mtDNA test will reveal information about just *one* person out of those 1,024 ancestors.

However, the authors of these articles fail to understand the incremental nature of genealogical research. Most genealogists spend a great deal of resources—both time and money—attempting to uncover even the smallest piece of information about single individuals within their family tree. Further, as we'll see in later chapters, being able to focus on one ancestor using DNA is incredibly valuable. The fact that the Y chromosome or mtDNA is found in just one of 1,024 ancestors at the tenth generation, for example, is part of what makes Y-DNA and mtDNA testing so powerful.

The authors of these articles also usually fail to understand atDNA testing, in which one DNA test examines many different lines of the family tree. Instead of obtaining information about just one ancestor in each generation, atDNA testing can potentially examine each of the many ancestors that provided DNA to our genomes. In the future, atDNA testing may even help identify ancestors who failed to provide DNA to us. (In other words, ancestors who are part of our genealogical family tree but not part of our genetic family tree; we'll talk more about this in chapter 6.) This ability to examine multiple ancestors with a single test also makes atDNA testing more challenging, but that is part of genetic genealogy's fun!

MISCONCEPTION #5:
DNA testing will reveal my health information.

This misconception has its grounding in some truth. One of the driving forces behind sequencing the first human genome was to use the information to understand the causes of disease and to find cures or treatments, so most genetic testing was done for medical reasons before the advent of genetic genealogy and personal genomics. As a result, it's not surprising that people anticipate the results of a genetic genealogy test will reveal their health information to both themselves and the testing company.

Indeed, there's no question that a genetic genealogy test can reveal health information about the test-taker. The genetic genealogy testing company 23andMe, for example, tests hundreds or thousands of locations in the genome that can be health informative, then provides that information to test-takers. Additionally, some DNA testing may inadvertently reveal health information, perhaps because a new scientific discovery uncovers a previously unknown health-related implication of the small percentage of the genome analyzed by genetic genealogy testing. Or the analysis can uncover one of a very few rare conditions that are already known to be related to the test-taker's health or other medical condition. For example, sequencing a commonly tested region (or **marker**) on the Y chromosome, DYS464, can reveal a serious deletion of a chromosome segment that results in male infertility. This deletion is very rare, occurring once in every four to eight thousand men. Some metabolic diseases can also be detected by full mtDNA sequencing.

However, you have several reasons not to be concerned about the possibility of revealing health information. While at one time, it was thought that sequencing a person's DNA would reveal what illnesses they would contract in their lifetime, current DNA testing simply can't lead to such dramatic conclusions. (Indeed, the 1997 movie *Gattaca* examined this now-erroneous prediction, with its main character, a man in the not-so-distant future with a DNA sequence that determines a short life expectancy, having to fight against the genetic-based discrimination he faces.) Scientists have discovered that the correlation between health and genetics is complex and that the environment plays a much larger role in determining our health. With the rare exception of individuals with serious genetic diseases that were already diagnosed before a genetic genealogy test, a test cannot reveal our major illnesses or eventual cause of death.

Further, with the exception of 23andMe, which intentionally offers health information as part of its test, most of the major genetic testing companies intentionally do not test health-related locations of the genome. And even if they do test for these locations, they scrub that information from the test results and do not provide it to the test-taker.

Accordingly, as this misconception has a strong basis in reality, test-takers should be aware of what they may learn about themselves before agreeing to take a test. Even though the correlation between our health and DNA is weak (and that even if armed with your entire DNA sequence of six billion nucleotides, a scientist can almost never predict your health, major illnesses, or cause of death), test-takers who remain worried about privacy and health can further alleviate their concerns by only testing at a company that intentionally does not provide health information.

MISCONCEPTION #6: **My parents and grandparents are deceased, so genetic genealogy won't help me.**

Although the ability to test parents and grandparents is invaluable, it's not necessary to successfully use DNA, and you can take advantage of some workarounds. Genetic genealogy is based on using DNA from today to understand and uncover the mysteries of yesterday; the DNA you walk around with today, which you inherited from your parents and grandparents, can be used to study your genetic family tree without testing any other relatives. Accordingly, don't despair if the answer you seek requires DNA from someone other than yourself. In most cases, DNA can be found in other living people who can be identified using traditional genealogical methods.

For example, males carry the Y-DNA that was given to them by their fathers, who received it from their fathers, and so on. As a result, there's usually no need to use Grandpa's Y-DNA when you predict your (or a living male relative's) Y-DNA to be the same. But testing can be more difficult if you're looking for Y-DNA from someone in your family tree other than your direct paternal line. To obtain that Y-DNA, trace the ancestor's male descendants and find a living direct male descendant willing to take a Y-DNA test. Similarly, first cousins received a huge amount of DNA from the shared grandparents, and second cousins received a significant amount of DNA from the shared great-grandparents.

The more family members you test, the easier it usually is to make discoveries and breakthroughs. At every generation, 50 percent of the previous generation can be lost if only one person tests. For example, you only carry half of your father's DNA and half of your mother's DNA, and on average you only carry 25 percent of the DNA from each of your four grandparents. If you can test your parents or grandparents, you can regain the 50 or 75 percent that might otherwise be lost. Similarly, testing aunts/uncles and siblings will regain an additional (though lesser) percentage since they inherited some of the same DNA and some unique DNA from your shared ancestors. We'll learn much more about this in chapter 6.

MISCONCEPTION #7:
Genetic genealogy testing is a violation of privacy.

Many people choose not to undergo DNA testing because they are afraid the results could be utilized for nefarious purposes, including by insurance companies and law enforcement agencies. Although the likelihood that your DNA will be used for an unintended purpose is extremely low, every genealogist must consider the implications of DNA testing before purchasing or taking a genetic genealogy test.

There is no question that we lose some control over our genetic information when we send away a saliva sample, although the three major DNA testing companies go to great lengths to project the genetic information in their databases. The pertinent question, therefore, is what a loss of control could potentially mean.

As we just learned, the correlation between our health and DNA is weak for most test-takers, and so you needn't worry about having sensitive information about your health revealed through DNA. For the most part, your DNA test results will simply reveal that you need to eat better and exercise more—advice you have probably heard already. Although specialized DNA can reveal more serious diseases for a very small number of people, most commercial genetic genealogy testing is designed to avoid this information.

Additionally, US federal law grants some limited protection in the form of the Genetic Information Nondiscrimination Act of 2008 (or GINA). GINA prohibits the use of genetic information by employers (who have fifteen or more employees) to make hiring, firing, or promotion decisions, and prevents health insurers from using genetic information to deny coverage or increase premiums. GINA, however, is not an absolute bar, and thus genetic information could potentially still be utilized by entities offering life insurance, disability insurance, and long-term care insurance.

Additionally, while it is possible that law enforcement agencies could potentially obtain your DNA results from a testing company, it is unlikely. There is no chain of custody for DNA test results from a commercial genetic genealogy company, meaning that the law enforcement agency cannot reliably ascertain that the DNA came from you. Subpoenaing this information from a testing company is also an expensive and complicated mechanism by which to obtain your DNA. We leave a trail of DNA everywhere we go, and thus it is far easier and cheaper for an agency to analyze a cup you leave behind at a restaurant or a bag of garbage you leave at the curb than to obtain a sample from a private testing company.

Having considered this, many of the concerns you might have about the privacy of your information is unfounded. While sending away a DNA sample to a testing company necessarily relinquishes some control over your DNA sequence, you're unlikely to face negative consequences from giving up that control.

MISCONCEPTION #8:
Because my mother/father/sibling shares atDNA with a genetic match, I should also share atDNA with that person.

Unless you understand how DNA is transmitted from one generation to the next, it's easy to believe that you share all of your parent's or sibling's matches. If my father shares DNA with his fourth cousin, shouldn't I share DNA with that same cousin (my fourth cousin once removed)? And if I don't, does that mean I'm actually not my father's child?

The answer depends on the genealogical relationship. As we'll see in chapter 6 on atDNA, the likelihood of sharing DNA with genealogical relatives is very high for close matches and very low for distant matches. The table below provides estimates from 23andMe, AncestryDNA, and Family Tree DNA <www.familytreedna.com> of the likelihood that genealogical relatives will share a detectable amount of DNA in common.

Relationship	23andMe	AncestryDNA	Family Tree DNA
Closer than second cousin	100%	100%	>99%
Second cousin	>99%	100%	>99%
Third cousin	~90%	98%	>90%
Fourth cousin	~45%	71%	>50%
Fifth cousin	~15%	32%	>10%
Sixth cousin or greater	<5%	<11%	<2%

While your father may have had a 45-percent chance of sharing DNA with that fourth cousin, your likelihood of sharing DNA with that same cousin—who is your fourth cousin once removed—is significantly less. However, if your parent's or sibling's genetic match is predicted to be very close, such as a first cousin, you should still share DNA with that genetic match.

This relates back to the concept of a genealogical family tree and a genetic family tree. Unless your sibling is an identical twin with the same DNA, your genetic family tree and your sibling's genetic family tree will be only partially overlapping. Your sibling will have some ancestors in his or her genetic family tree—and hence some genetic cousins and matches—who you don't have, and vice versa.

The same is true for your parent's genetic family tree. However, since you only inherited 50 percent of your parents' DNA, you are a *subset* of each parent's genetic family tree. As a result, you must share each and every one of your true genetic matches with one (or both) of your parents, but your parents do not need to share each of their genetic matches with you.

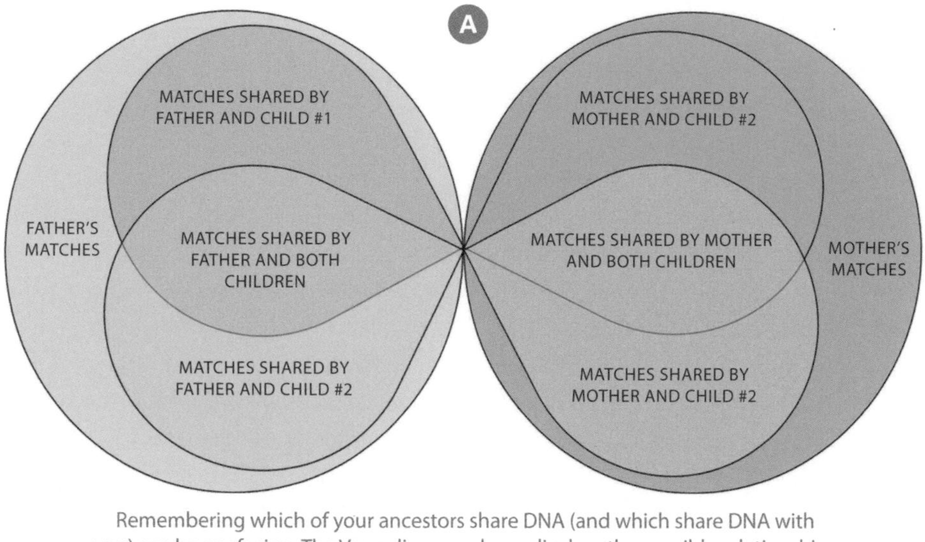

Remembering which of your ancestors share DNA (and which share DNA with you) can be confusing. The Venn diagram above displays the possible relationships between a father's and mother's lines and that of their children.

Image A shows the overlap of matches between two parents and their two children. Both the mother and father have matches who each of the children don't have. Each child's universe of possible genetic matches, however, is entirely within the father and mother's match lists. The children share many of their matches in common ("Matches shared by father and both children" and "Matches shared by mother and both children") but each share matches with a parent that the other does not share.

In many cases, the mother and father will share genetic matches as well, but the parents are shown as not sharing any DNA or genetic matches for the purposes of this diagram.

MISCONCEPTION #9: **I should share DNA with my genealogical relatives.**

This misconception is closely related to the previous misconception. Many test-takers purchase a DNA test expecting to receive a list of all *genealogical* relatives who have also taken a DNA test. However, because each generation receives only 50 percent of the DNA from the previous generation, you will actually fail to match most of your genealogical cousins, at least beyond about the fourth-cousin level. As the previous table shows, while you will match all your second cousins and closer relatives, the likelihood of matching fourth cousins and beyond gets exceedingly rare with each generation.

This does not mean, however, that you and a fourth cousin who does not share DNA with you can't both have the same common ancestor in your genetic family trees. In order to share DNA with a genealogical relative, *all* of the following conditions must be met:

1. You inherited DNA from a certain ancestor.
2. Your genealogical cousin inherited DNA from that same ancestor.
3. You and your genealogical cousin inherited at least some of the same DNA from that shared ancestor.

If you and a genealogical cousin fail to share DNA in common, one or more of these three conditions have not been met. Another condition is that the shared segment(s) of DNA must be detectable, meaning it must be a sizeable enough segment of DNA to be identified by the testing company. We'll talk more about company thresholds in chapter 6.

Understanding who you might share DNA with—and the many reasons you might not share DNA with them—is one of the most important aspects of genetic genealogy.

MISCONCEPTION #10: **My ethnicity estimate from the testing company should match my known genealogy.**

The ability to predict ethnicity based on the genealogical family tree is one of the biggest misconceptions in genetic genealogy. It is also one of the biggest complaints from test-takers, confusing and angering people who expect their ethnicity estimate to perfectly match their known genealogy.

However, it is impossible to predict an ethnicity estimate based on known genealogy for a number of reasons. First, as we'll see in chapter 9, ethnicity estimates are inherently limited by several factors including the size and composition of the **reference populations**—the populations from all over the world to which every test-taker is compared—used for the analysis. The testing companies are continually adding to their reference populations, but they are still rather small. As a result of these factors, an ethnicity *estimate* is simply an estimate and thus should not be considered an absolute or final determination. Indeed, it should be expected that every ethnicity estimate will change at least slightly over time as reference populations continue to grow and the testing companies improve their ethnicity estimate algorithms.

In addition to the inherent limitations of ethnicity estimates, it's impossible to predict an individual's ethnicity due to the limited knowledge most people have about their genetic family tree. As we saw in the previous chapter, everyone has both a genealogical family tree and a genetic family tree, with the genetic tree being a small subset of the

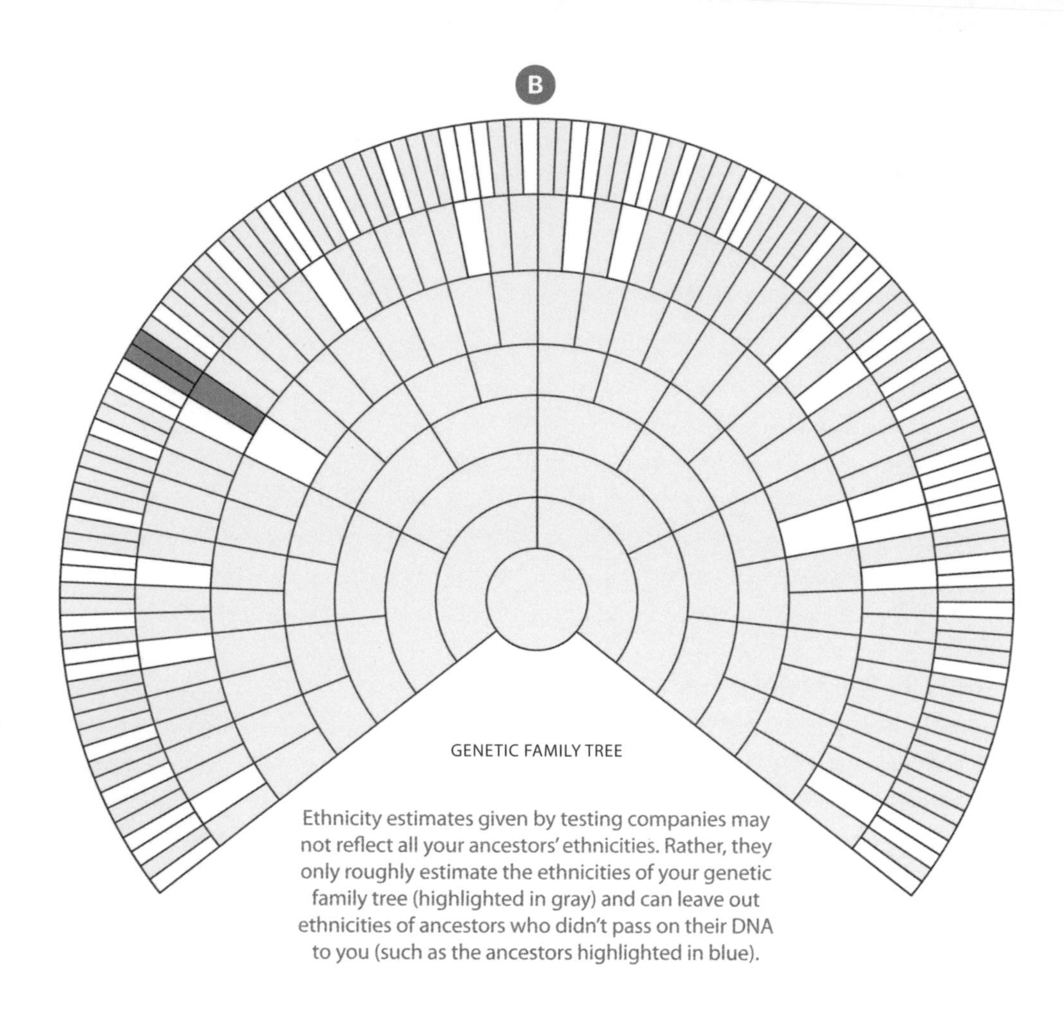

Ethnicity estimates given by testing companies may not reflect all your ancestors' ethnicities. Rather, they only roughly estimate the ethnicities of your genetic family tree (highlighted in gray) and can leave out ethnicities of ancestors who didn't pass on their DNA to you (such as the ancestors highlighted in blue).

genealogical tree. However, the issue for ethnicity estimates is that you don't know *which* subset of your genealogical tree makes up your genetic tree.

For example, image B shows a genetic family tree with a particular ethnicity of interest highlighted in blue. However, the individuals in the tree who actually provided DNA to the test-taker are highlighted in gray. Since the individuals highlighted in blue failed to provide any DNA to this particular descendant, their ethnicity cannot be detected using only the descendant's test results. And since this isn't on the test-taker's Y-DNA or mtDNA line, those tests will not detect this ethnicity either. There is no question that this ethnicity existed in the test-taker's family tree, but it cannot be detected with the test-taker's current test results.

This phenomenon will occur all across the test-taker's family tree, meaning that the test-taker cannot predict which ancestors' ethnicities might be detected. Further, as we'll see in later chapters, even for ancestors who are located within the test-taker's genetic family tree, such little DNA is passed down after a few generations and that some ethnicities may not be detected.

In some instances, the test-taker may have a family tree with roots in a single location for hundreds of years, such as in England or continental Europe. In this case, it is often possible to have a good approximation of this ethnicity estimate, although even these populations cannot exactly predict the ethnicities of the genetic family tree. Most people come from regions of the world where populations have not been completely stable and weren't isolated for the hundreds or thousands of years it takes for ethnicity to be well defined.

MISCONCEPTION #11:
The relationship prediction provided by the testing company is the actual genealogical relationship.

Each of the testing companies provides a relationship prediction based largely on the amount of DNA that the test-taker shares with the genetic match. The relationship predictions are usually a range of possible relationships rather than an exact relationship prediction. Each of the major genetic genealogy testing companies has a slightly different set of relationship predictions. Family Tree DNA provides a relationship range

C

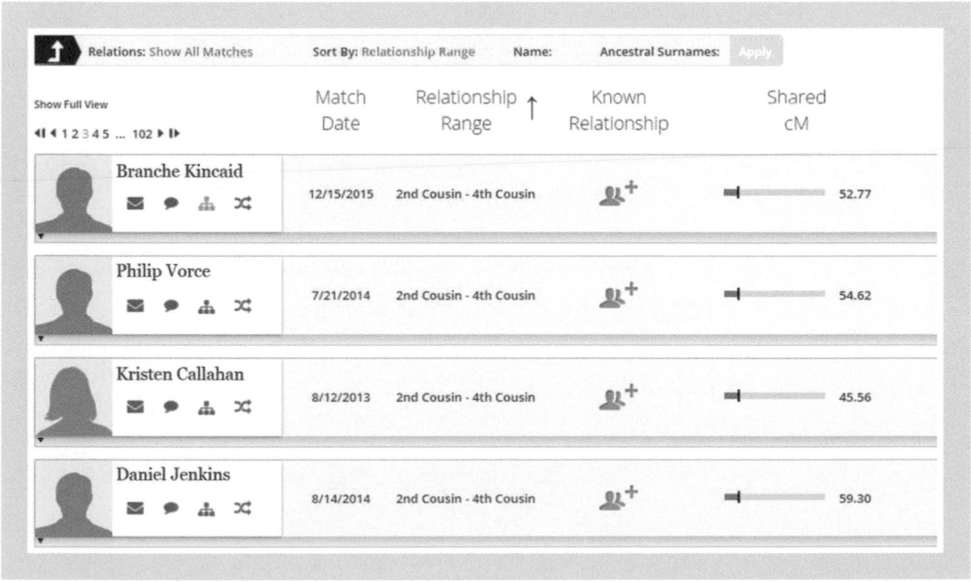

Relations: Show All Matches   Sort By: Relationship Range   Name: Ancestral Surnames: Apply

Show Full View

	Match Date	Relationship Range	Known Relationship	Shared cM
Branche Kincaid	12/15/2015	2nd Cousin - 4th Cousin		52.77
Philip Vorce	7/21/2014	2nd Cousin - 4th Cousin		54.62
Kristen Callahan	8/12/2013	2nd Cousin - 4th Cousin		45.56
Daniel Jenkins	8/14/2014	2nd Cousin - 4th Cousin		59.30

Family Tree DNA provides relationship ranges for DNA matches, such as from second cousins to fourth cousins.

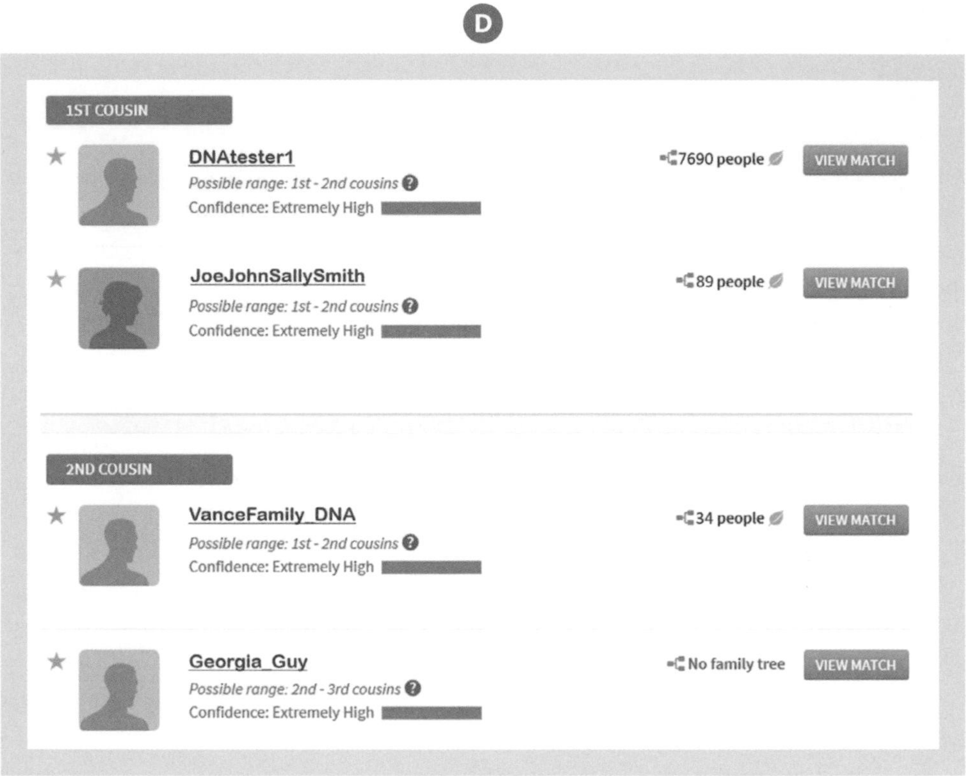

Like Family Tree DNA, AncestryDNA provides a relationship range, as well as a confidence interval and a quick link to the match's family tree.

(image C), such as a range from second cousins to fourth cousins. AncestryDNA arranges matches into categories such as "1st Cousin" and "2nd Cousin," but also provides a possible relationship range (image D). Clicking on the question mark next to each relationship range produces a pop-up with additional information. Testing company 23andMe also provides relationship ranges. In image E, the relationships range from second cousin to sixth cousin. Clicking on a match reveals a profile page with a more specific relationship prediction.

We'll also learn a lot more about these relationship predictions in chapter 6.

Accordingly, the testing companies provide a relationship prediction, but the test-taker is not guaranteed that the prediction is the *exact* genealogical relationship between those individuals. Similar relationships can result in similar amounts of DNA being shared by genetic matches, complicating relationship predictions. For example, a first cousin and a first cousin once removed may share similar amounts of DNA with a relative.

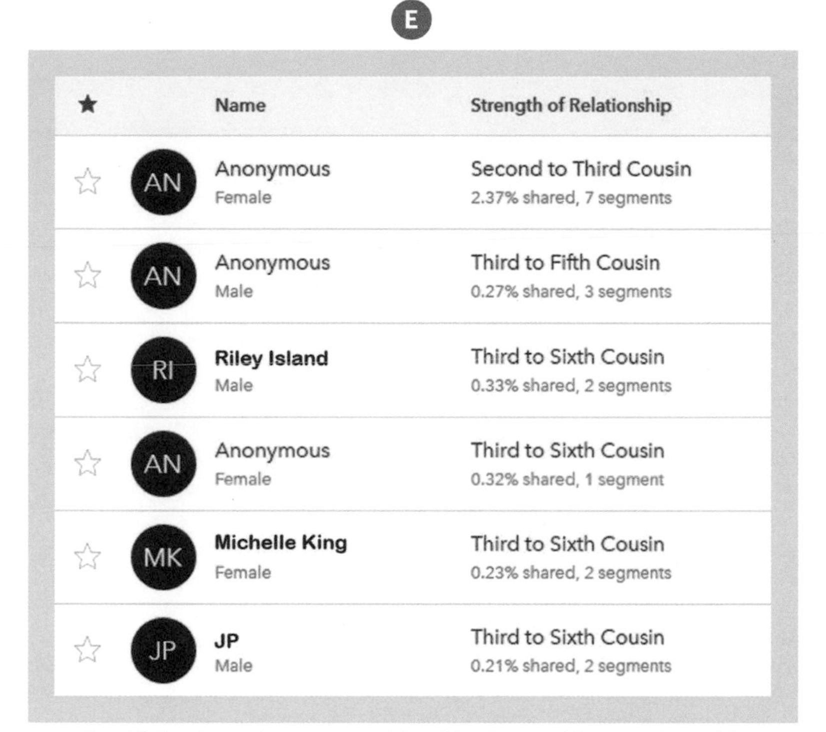

E
Name
Strength of Relationship
Anonymous
Female
Second to Third Cousin
2.37% shared, 7 segments
Anonymous
Male
Third to Fifth Cousin
0.27% shared, 3 segments
Riley Island
Male
Third to Sixth Cousin
0.33% shared, 2 segments
Anonymous
Female
Third to Sixth Cousin
0.32% shared, 1 segment
Michelle King
Female
Third to Sixth Cousin
0.23% shared, 2 segments
JP
Male
Third to Sixth Cousin
0.21% shared, 2 segments

The third major testing company, 23andMe, also provides test-takers with relationship range estimates for potential matches, along with a percentage of shared tested DNA.

Further, relationship predictions are complicated when there are multiple relationships within the family tree. As just one example, double first cousins (relatives who share both sets of grandparents in common; e.g., if a pair of brothers has children with a pair of sisters) can share an amount of DNA very similar to half-siblings. Other, more distant relationships can also impact the predictions. Additionally, relationship predictions cannot be accurate at more distant genealogical relationships. For example, a seventh cousin and a tenth cousin typically share very small but possibly similar amounts of DNA with their relatives.

Even with these limitations and misconceptions, genetic genealogy can be an informative and exciting addition to traditional research that can often be used to answer specific genealogical mysteries. There are many—even hundreds—of genealogical success stories thanks to the proper use of genetic genealogy. To read about just a few of these inspiring success stories, see the International Society of Genetic Genealogy (ISOGG) Success Stories <www.isogg.org/successstories.htm>. In part two of this book, we'll dive into the different types of DNA testing and how they may be used to advance your genealogical research.

CORE CONCEPTS: COMMON MISCONCEPTIONS

- Genetic genealogy can be entertaining, but it is also an essential research tool for every genealogist.

- Anyone can take an atDNA or mtDNA test. Although only men can take a Y-DNA test, women can find a male relative to take a Y-DNA test for their research.

- DNA testing, alone, does not provide the test-taker with a completed family tree. Although mtDNA and Y-DNA tests examine just one person in each generation, that is part of what makes them such powerful research tools.

- DNA testing can potentially reveal health information, but understanding the different company offerings and how the tests can reveal health information can help prevent most issues.

- Although testing parents and grandparents can be useful, those relatives are not necessary to utilize genetic genealogy.

- Since your genetic family tree is just a small subset of your genealogical family tree, you will not share DNA with all of your genealogical relatives. In fact, you will not share all of your parent's or sibling's genetic matches, as you inherit only 50 percent of your parent's DNA, and you only share about 50 percent of your DNA with your sibling.

- You cannot accurately predict your ethnicity estimate based on your known genealogy. Similarly, relationship predictions from testing companies are just estimates and should only be considered as possible relationships requiring additional research.

3

Ethics and Genetic Genealogy

Should you tell your uncle that he doesn't share DNA with his sister (your mother)? Does it make a difference if that uncle is 35 or 95 years old? Should you help an adoptee who is predicted to be your first cousin, meaning that one of your aunts or uncles had a child you don't know about, or should you ignore her requests?

Since DNA can both reveal unexpected biological relationships and disprove expected relationships, DNA testing can raise many ethical issues. Not every test-taker will encounter these tough questions, but as more people are tested, more opportunities for these issues arise. Making test-takers aware of the possible outcomes of DNA testing (as we've done throughout this book) can help prevent some of these issues, but this chapter will discuss how to work through these difficult questions that result from DNA testing and provide some ethical frameworks to consider when making decisions.

Ethical Issues Encountered by Test-Takers

If you or a relative are considering DNA testing, what are some of the ethical issues you might face? How should you deal with these issues once you discover them? Before we discuss how to anticipate or deal with any ethical issues, we will first take a brief look at a few of the ethical issues or dilemmas that might be encountered by a test-taker or genealogist as a result of DNA testing. There are many more, with others just being discovered as DNA testing becomes increasingly powerful.

Relationship Disruption or Discovery

Genetic testing can result in the discovery of genealogical relationships that the test-taker didn't know existed. It is commonplace to find new second or third cousins as a result of DNA testing, since these relationships are more distant and families often lose track of each other after only a few generations. However, finding a new relative at the first-cousin level (or even closer) can be unexpected. For example, it is not uncommon to find new half-siblings, aunts/uncles/nieces/nephews, first cousins, or other previously unknown close relatives through test results. These relatives can be the result of a wide variety of family situations and can be a complete surprise to everyone involved or perhaps a family secret that was widely known but not discussed.

Genetic testing can also disrupt a genealogical relationship that the test-taker thought was based on a genetic relationship. In this common scenario, the test-taker and a relative—usually a second cousin or closer—both take a DNA test and discover they don't share DNA as expected. The two relatives may discover they share absolutely no DNA in common. "Misattributed parentage" is the term given to the situation where a DNA test result suggests that a relationship is inaccurate, most commonly in Y-chromosomal DNA (Y-DNA) testing or when doing autosomal-DNA (atDNA) testing of close relatives (second cousin or closer). Misattributed parentage is much harder to detect if the relationship is further back in time.

Several ethical questions can arise when a DNA test unexpectedly reveals or disrupts a genealogical relationship. For example, the most logical question that may arise is whether you should share this information with the relative(s) that it impacts. If you ask a first cousin to take a DNA test and he doesn't share DNA with you—meaning that he is not actually your first cousin—do you share that information with him, his parents, or your parents? As we'll discuss in a later section, you have an obligation to share his results with him, but no obligation to explain the results or ensure that he understands them. Is it ethical to keep this information as secret as possible? Sensible, ethical people will possibly disagree on the proper course of action in this scenario, and there are no laws that require a specific response.

As another example, say you receive a request for assistance from a new close relative who was adopted (such as a predicted first cousin). The logical question that will arise is whether you should assist this new first cousin by sharing information with her. Should you share information about your family that the person could use to potentially identify her biological parent (who will be your aunt or uncle), should you ignore her request, or should you ask her not to investigate this connection? Your response to the new relative scenario might be different if, for example, the new relative appears to be a maternal half-uncle, and your maternal grandparents (one of whom appears to have been his parent) are deceased. In this scenario, fewer of the "key players" are involved and the revelation of a new family member may have less of an impact. Once again, sensible and ethical people will disagree on the proper course of action here.

Unfortunately, most test-takers do not understand prior to testing that the results can discover or disrupt long-held relationships. Other test-takers ignore the possibility because they are confident in their genealogical hypotheses. However, test-takers should be prepared to handle scenarios in which they discover either new relatives or that their existing "relatives" aren't actually biologically related.

Adoption

Adoptees are one of the largest communities to embrace genetic genealogy testing, as it often provides the ability to circumvent stringent state laws regulating adoption records. As a result, test-takers may find they are closely related to an adoptee about whom they may or may not have had prior knowledge.

Discovering an adoption can lead to many difficult ethical issues. For example, what are the test-taker's obligations to the adoptee? What are the test-taker's obligations to his or her own family? Should adoptees be able to circumvent state laws that were enacted to restrict access to the facts surrounding an adoption? Whose rights should prevail: the adoptee's, the biological parents', or the adoptive parents'? Everyone should have a right to their own DNA, regardless of who provided that DNA. Regardless, navigating this minefield of potential ethical issues can be difficult.

Donor Conception

In the decades since sperm and egg donation first became possible, donors were promised anonymity, if they so desired. Many donors relied on that promise when they decided to go forward with the donation. However, the resulting offspring can circumvent that anonymity for less than a hundred dollars with the purchase of a DNA test.

Although every child deserves to have that information about his or her genetic heritage, this clearly conflicts with the anonymity that was promised to the egg or sperm donor. Unfortunately, the only way to protect the anonymity of donors is to completely prevent any and all DNA testing, an action that would have far more damaging

GENETIC EXCEPTIONALISM

DNA testing is a genealogical source that provides the test-taker with information about one or more genealogical relationships, which by their nature are personal and often sensitive. But do the results of DNA testing merit different treatment than conclusions drawn from other types of genealogical records?

Genetic exceptionalism is the belief that genetic information is unique and needs to be treated in a different way from other genealogical information. Proponents of genetic exceptionalism believe that genetic information requires strict and explicit protection, in part because of the ability of DNA to reveal information not just about the individual, but also about the individual's family. In addition, DNA is in some ways predictive, as it can indicate predisposition towards (or even the presence of) certain genetic and/or medical maladies, some of which can be dangerous.

There is no question that genetic testing can reveal family secrets both new and old. Indeed, tens of thousands of genetic genealogists purchase DNA testing for exactly that reason: to discover the truth behind their own family secrets. Many other customers of genetic genealogy testing learn about family secrets they never knew existed. Some people will be thrilled to learn the truth about these family secrets, while others might be devastated. And with genetic testing becoming increasingly prevalent, these secrets are being revealed at an incredible rate.

But does DNA really reveal that much more information about families than traditional genealogy research? Other types of genealogical records, such as census records, birth certificates, deeds, or tax records often provide similar information for genealogists. For example, a birth certificate can reveal that parents who raised a child were in fact not its biological parents, or a census record can reveal that a family was in fact a blended family due to the census-taker's use of words like "half" or "step." After all, unexpected pregnancies, infidelity, adoption, divorce, and other family events that can trigger emotional responses are not experiences found only in modern times.

Even without DNA testing, genealogists may be able to uncover potentially sensitive information about relatives. A descendant of Helen Bulen, born about 1889 in New York, for example, might be surprised to discover her age from a census record. Although family relationships were not recorded in the 1892 New York state census, families were usually accounted for together. In this record, three-year-old Helen Bulen is living with Frank Bulen (age fifty-three) and Helen Bulen (age sixty-two). Although Helen would later identify Frank and Helen as her parents in a Social Security application, clearly Helen Bulen could have not given birth to Helen in 1889 at approximately age fifty-nine.

Similarly, a descendant or relative of Leander Herth might be surprised to discover in the 1900 census that Leander was a "Foundling" boy born in June 1898. Although Leander was a foundling, he possessed the surname of the family he was living with in the census, and it would be easy for a descendant to believe that he was a biological child of Joseph and Emma Herth. Many

The detail "Foundling" in Leander Herth's entry describes a family situation that's more complicated than a genealogist may have originally believed, especially since Leander went by the surname Herth.

relationships and events are lost through time, either intentionally or inadvertently.

In addition to traditional records used for genealogical research, adoptees have been pushing for access to sealed adoption records. These records, perhaps more than any other, contain direct evidence of non-biological family relationships. In 2010, for example, the state of Illinois passed a law giving adoptees over age twenty-one the right to request a copy of their original birth certificates. Since the law went into effect, the state has issued more than ten thousand birth certificates to adult adoptees according to a 2014 article by the *Chicago Daily Law Bulletin* <**www.chicagolawbulletin.com/Archives/2014/05/22/Adoption-Birth-Certificate-5-22.aspx**>. Other states have enacted or are considering enacting similar laws. These laws have revealed thousands of non-biological family relationships without involving DNA. Accordingly, DNA does not have a monopoly on revealing family secrets.

Although genetic exceptionalism has many proponents, particularly in academic circles, the theory has been soundly rejected by many genetic genealogists who work with all different types of genealogical records on a daily basis. Given the kind of information available to genealogists through other types of records, it seems illogical to object to DNA testing because it reveals information about the individual and an individual's family without also objecting to all forms of traditional genealogical research and laws that open records to adoptees. All records have the potential to reveal information about non-biological family relationships, and the kind of family events that can be turned up by modern DNA research and trigger emotional responses from descendants—adoptions, miscarriages, infidelity, divorce, etc.—aren't unique to modern times.

Although DNA is capable of revealing information about a test-taker and the test-taker's relatives and ancestors, it is just one of many such record types. Genealogists use a multitude of different record types to recover and rebuild information about both biological and non-biological relationships of the past and present. Just as genealogists must be careful about revealing information discovered about living people in census records or other record types, genealogists must be careful with the information revealed about living people by DNA testing.

consequences. Despite the ability of DNA testing to reveal information about egg and sperm donors, potential donors are sometimes promised anonymity even today.

Privacy

Privacy is a major concern for genetic genealogists. For example, DNA test results can implicate not just the test-taker, but also the test-taker's close family members and even the test-taker's distant genetic relatives. Given these far-reaching consequences, should a test-taker be able to take a DNA test without the permission of close relatives? Should a test-taker publicly share information about matches without the permission of everyone in the list? These privacy issues are among the most prevalent in the genetic genealogy community.

Preventing and Resolving Ethical Issues

Learning more about DNA tests and their results is the most effective way for test-takers to prevent and deal with the ethical issues raised by testing. As a result, genealogists must have an in-depth understanding of the possible ethical issues in order to educate themselves and other test-takers.

As genealogist Debbie Parker Wayne wrote in the *Association of Professional Genealogists Quarterly*, most genealogists "believe that handling genetic information in the same way we handle genealogical information gathered from documents is the best path—that 'genetic exceptionalism' is not a valid theory for genealogy, even if it may have medical applications." However, genealogists who are handling genetic information potentially full of family secrets have limited guidance on how to deal with this sensitive information.

Since DNA is not unique in its ability to reveal unknown, secret, or forgotten genealogical information, genealogists should see how their colleagues have dealt with privacy and ethical issues in other areas of genealogical research. For example, the National Genealogical Society's Standards for Sharing Information with Others **<www.ngsgenealogy.org/cs/standards_for_sharing_information>**, written in 2000, advises genealogists to "respect the restrictions on sharing information that arise from the rights of another ... as a living private person," and "require some evidence of consent before assuming that living people are agreeable to further sharing of information about themselves."

Similarly, the Code of Ethics of the Board for Certification of Genealogists **<www.bcgcertification.org/aboutbcg/code.html>**, which only regulates board-certified genealogists but provides insight into this issue, requires that these genealogists "...keep confidential any personal or genealogical information given to [them], unless [they] receive written consent to the contrary."

Arguably, these broad standards and ethical guidelines provide enough guidance for genetic genealogists. However, they do not *specifically* address the ethical issues that can arise from DNA testing. Indeed, in the December 2013 issue of the *National Genealogical Society Quarterly*, editors Melinde Lutz Byrne and Thomas W. Jones lamented the lack of standards for using the results of DNA testing. As they noted, "[a]s difficult as it is to cite, describe, explain, or utilize this rapidly evolving tool, the real DNA-test quagmire is ethical."

The Genetic Genealogy Standards

Recognizing this lack of guidance, a group of genealogists and scientists came together in the fall of 2013 to draft standards for DNA testing. Over the course of the next year, this group drafted a document called the **Genetic Genealogy Standards**, which were officially released on January 10, 2015 as an invited paper at the Salt Lake Institute of Genealogy Colloquium **<www.thegeneticgenealogist.com/2015/01/10/announcing-genetic-genealogy-standards>**. A copy of the Genetic Genealogy Standards is available at **<www.geneticgenealogystandards.com>**.

The Genetic Genealogy Standards are directed to "genealogists," which is defined in the Standards as anyone who takes a genetic genealogy test, as well as anyone who advises a client, family member, or other individual regarding genetic genealogy testing. As a result, the Standards are directed at consumers rather than genetic genealogy testing companies.

The Standards are divided into two sections: The first section is directed to standards for obtaining and communicating the results of DNA testing, and the second section is directed to standards for interpreting DNA test results.

There are no absolutely right answers or absolutely wrong answers when it comes to ethical questions raised by genetic genealogy testing. However, the Genetic Genealogy Standards were written to help provide some guidance in preventing and responding to these ethical questions. In this section, I'll outline some of the most important points presented by the Standards, together with some of the reasoning behind them.

STANDARD #1: **Company Offerings**

> "Genealogists review and understand the different DNA testing products and tools offered by the available testing companies, and prior to testing determine which company or companies are capable of achieving the genealogist's goal(s)."

This mandates that genealogists have at least a basic understanding of the various types of DNA testing and what the testing companies offer. DNA testing can be expensive, especially when testing at several different companies.

Accordingly, it is important that, in order to maximize our testing dollars, or the testing dollars of the relatives we ask to test, we ensure that the tests we order are capable of achieving our goals. We should not, for example, order an mtDNA test to examine the paternal line of a family tree (which, as we'll see in chapter 5, will more likely benefit from Y-DNA testing).

STANDARD #2: **Testing with Consent**

> "Genealogists only obtain DNA for testing after receiving consent, written or oral, from the tester. In the case of a deceased individual, consent can be obtained from a legal representative. In the case of a minor, consent can be given by a parent or legal guardian of the minor. However, genealogists do not obtain DNA from someone who refuses to undergo testing."

The ethical concerns of genetic genealogy testing has never been more clear than in a 2007 *New York Times* article describing the great lengths that genealogists would go to obtain DNA from relatives, often without consent **<www.nytimes.com/2007/04/02/us/02dna.html>**. As profiled in the article, some genealogists essentially "stalk" potential relatives to get their DNA information, resorting even to salvaging a needed test subject's coffee cup from a garbage can.

Under the Standards, however, any effort made without consent, either from the provider, a legal representative, or a parent/legal guardian, is prohibited. The only exception is situations where DNA testing is specifically mandated by law or court order. For example, some genealogists are routinely involved in cases in which an individual refuses to test but is forced to do so pursuant to a court order.

STANDARD #3: **Raw Data**

> "Genealogists believe that testers have an inalienable right to their own DNA test results and raw data, even if someone other than the tester purchased the DNA test."

This standard also advises that a genealogist must make raw data available to the person who provided the DNA sample. For example, if a genealogist purchases a test for an aunt, the genealogist must make the raw data available to the aunt even though she didn't purchase the test. This promotes openness and sharing among those who purchase tests and those who provide the DNA that is analyzed by the test. This also reinforces the belief of many genetic genealogists that individuals have a right to their own genetic heritage.

Fortunately, the three major testing companies make raw data—e.g., a *GG* result at location rs13060385 on chromosome 3—available to the test-taker, so testing or recommending testing at these companies falls within the guidelines of the Standards. The

Standards do not, however, specifically address a scenario in which a genealogist recommends a test at a company that doesn't return raw data. Would recommending such a test violate the Standards?

STANDARD #4: **DNA Storage**

> "Genealogists are aware of the DNA storage options offered by testing companies, and consider the implications of storing versus not storing DNA samples for future testing. Advantages of storing DNA samples include reducing costs associated with future testing and/or preserving DNA that can no longer be obtained from an individual. However, genealogists are aware that no company can guarantee that stored DNA will be of sufficient quantity or quality to perform additional testing. Genealogists also understand that a testing company may change its storage policy without notice to the tester."

Formulating an efficient DNA testing plan capable of addressing a research goal is an essential component of responsible and informed genealogical research. Often this research plan will involve decisions about current and future DNA testing. As we'll discuss later in the book, only Family Tree DNA currently offers the ability to order a DNA test using a stored sample remaining from a previous test or sample collection. Accordingly, if future testing—such as an upgrade or a different type of test—is an option, then storage options must be considered. This may involve, for example, only testing at Family Tree DNA, or testing at multiple companies including Family Tree DNA. Understanding all of the testing strategies and storage options available is an essential facet of genetic genealogy testing.

STANDARD #5: **Terms of Service**

> "Genealogists review and understand the terms and conditions to which the tester consents when purchasing a DNA test."

Unfortunately, consumers don't read terms of service agreements before they accept them, in part because of the time it would take to read them all, and in part because they are often written in barely comprehensible legalese. In 2005, the computer repair business PC Pitstop added a clause to its End-User License Agreement offering a large financial reward to whoever contacted them at a given e-mail address. Amazingly, it took three thousand sales and five months before the first person discovered and responded to the clause to claim the reward **<techtalk.pcpitstop.com/2012/06/12/it-pays-to-read-license-agreements-7-years-later>**. With DNA test results, it is important that genealogists read and understand the possible implications of testing prior to purchasing or recommending a test.

STANDARD #6: **Privacy**

"Genealogists only test with companies that respect and protect the privacy of testers. However, genealogists understand that complete anonymity of DNA tests results can never be guaranteed."

Once again, privacy is emphasized as an important aspect of DNA testing. Although it almost goes without saying that a genealogist shouldn't test with a company that doesn't protect a test-taker's privacy, the Standards committee felt this was too important to omit.

However, it is also vitally important that everyone who agrees to take a DNA test understand that no one can guarantee complete anonymity of DNA test results, even if a test-taker uses a pseudonym. Indeed, the goal of most DNA testing is to find genetic matches. It should come as no surprise, therefore, that the results of DNA testing can be used to identify the test-taker even if the test is anonymous or de-identified by the testing company.

STANDARD #7: **Access by Third Parties**

"Genealogists understand that once DNA test results are made publicly available, they can be freely accessed, copied, and analyzed by a third party without permission. For example, DNA test results published on a DNA project website are publicly available."

Once a test-taker agrees to make her DNA publicly available, there is no further protection for that DNA. Publishing results on the website of a surname project, for example, means that anyone is free to copy and use those results; there is likely no copyright protection in raw DNA data. Further, when someone accesses DNA test results on a public website with whom he has no contractual arrangement, there is no restriction on how those results can be utilized. As a result, it is important that genealogists understand the effect of making DNA test results publicly available.

The benefits of DNA testing must always be balanced with concerns about privacy for both the test-taker and the test-taker's genetic and genealogical family members. While the only way to keep DNA completely private is to avoid DNA testing—although even this is not an absolute guarantee—the power of DNA testing to reveal genealogical information cannot be realized if it is never tested.

STANDARD #8: **Sharing Results**

"Genealogists respect all limitations on reviewing and sharing DNA test results imposed at the request of the tester. For example, genealogists do not share or otherwise reveal DNA test results (beyond the tools offered by the testing company) or other personal information (name, address, or email) without the written or oral consent of the tester."

Asking relatives for their DNA can be challenging. People often have reasonable concerns about privacy and the misuse of test results that prevents them from testing. To alleviate their concerns, it is possible to offer the relative restrictions on sharing of their test results. For example, it is common to use a pseudonym or initials when testing a relative.

Once an individual has agreed to undergo testing, any restrictions she puts on that test must be honored unless the test-taker—or perhaps a legal representative or heir—is contacted to change the terms of the original agreement or arrangement. This is even true if the original restrictions hinder future research. For example, if a relative asked to use a pseudonym, the relative's name cannot be shared with genetic matches unless consent has been established. The relative's raw data or test results cannot be uploaded to a third-party site like GEDmatch <**www.gedmatch.com**>, for example, unless the relative's consent has been obtained.

The Standards don't address a situation in which a test-taker provided a genealogist consent to test DNA with certain restrictions, and later died. Does future consent have to be obtained from the heirs of the test-taker, or does the deceased test-taker no longer have any rights to the DNA? These questions remain to be answered, and may not have a direct correct or incorrect answer.

STANDARD #9: **Scholarship**

> "When lecturing or writing about genetic genealogy, genealogists respect the privacy of others. Genealogists privatize or redact the names of living genetic matches from presentations unless the genetic matches have given prior permission or made their results publicly available. Genealogists share DNA test results of living individuals in a work of scholarship only if the tester has given permission or has previously made those results publicly available. Genealogists may confidentially share an individual's DNA test results with an editor and/or peer-reviewer of a work of scholarship."

Genealogists have long recognized the importance of privatizing and redacting the names of living people. In the DNA realm, however, genetic genealogists have been slow to embrace privatization. It is far too easy to take a screenshot of results and innocently share it online or in a presentation without removing the names of matches.

Matches should only be revealed if they've given explicit permission. Although this can be impractical with a long list of matches, it is necessary to protect privacy. There may be a few exceptions this rule—for example, an author sharing test results with an editor or peer-reviewer in an unpublished work of scholarship.

One gray area, however, is when results are shared with just one other person or a small group of people. For example, if a genealogist looks over his results with a friend or family member, should he first somehow redact the list of genetic matches? Or what if a genealogist displays her results during a meeting of a small DNA special interest group? These questions are unresolved, but most genealogists believe there can be an adequate balance between privacy and exploring results with others.

STANDARD #10: **Health Information**

> "Genealogists understand that DNA tests may have medical implications."

It is easy to tell a test-taker that AncestryDNA and Family Tree DNA, for example, do not reveal health information about the test-taker (23andMe intentionally tests for and reveals health information). However, this is an inaccurate statement. As we will see in the chapters about Y-DNA and mitochondrial DNA (mtDNA), some tests can reveal health information. Not only can the results of Y-DNA, mtDNA, and atDNA testing be mined for information about certain known medical conditions, but these results may lead to more discoveries in years to come as scientists better understand DNA. An otherwise innocent result today may indicate a propensity for a trait or medical condition tomorrow.

Accordingly, the most a genealogist can promise test-takers is that a test through AncestryDNA and Family Tree DNA does not intentionally look for health information and usually does not reveal health information about the test-taker.

STANDARD #12: **Unexpected Results**

> "Genealogists understand that DNA test results, like traditional genealogical records, can reveal unexpected information about the tester and his or her immediate family, ancestors, and/or descendants. For example, both DNA test results and traditional genealogical records can reveal misattributed parentage, adoption, health information, previously unknown family members, and errors in well-researched family trees, among other unexpected outcomes."

Understanding *before testing* that DNA test results can disrupt existing relationships and uncover new ones can prevent many of the ethical issues that arise. As the size of the commercial DNA testing databases grows, it becomes progressively easier to detect and solve these revealed and disrupted relationships. This is a decidedly positive development for anyone trying to uncover their genetic heritage, but it can also raise ethical questions that we've discussed in detail in this chapter.

Applying Ethical Standards in Research

Now that we've examined some potential ethical issues that you may encounter through genetic genealogy testing (and the ethical standards encouraged by members of the genetic genealogy community), how should you behave ethically while doing research? Many people will travel through their DNA journey without experiencing any unexpected results, while others might encounter an ethical issue with their very first test results. This doesn't mean that people should be fearful of genetic genealogy. Rather it simply means that people must be aware of the possibilities, and prepared for these possibilities.

In practice, there are a series of simple steps that you can take to avoid or mitigate the impact of these ethical issues:

1. **Understand the possibilities.** Be aware of the possible outcomes of DNA testing, including the possibility of discovering new close relatives you didn't know existed, as well as the possibility of discovering that close relatives are in fact not related to you. This chapter explains many of these possibilities in detail.
2. **Study the Genetic Genealogy Standards** and ensure that your DNA testing plan meets or exceeds all of the Standards. For example, do you understand the terms and conditions of the testing company you've chosen to test with? When you get the test results back, do you plan to respect the privacy of your matches?
3. **Encourage others to be ethical** by sharing the Genetic Genealogy Standards when asking other people to test, and explain the risks of DNA testing. Discuss the possible outcomes with them, and ask whether they'd like to be informed if the results don't align with expectations. Although this will potentially result in people refusing to test, this is preferable to testing people who don't completely understand or appreciate the potential outcomes and therefore might have a negative experience with genetic genealogy testing.
4. **Respond to issues responsibly.** When an ethical issues arises, be ready for it and be able to respond to it discretely. If you've asked a relative whether they want to be informed of an unexpected result, much of your response will already be mapped out for you; either they won't want to be informed, or they will want to be informed and you'll have to craft a thoughtful way to do that.
5. **Be prepared for the unexpected.** Unfortunately, unexpected ethical issues can arise even with the best planning. For example, you could receive a request for assistance from a new genetic match predicted to be a new first cousin, meaning that one of your aunts or uncles—who haven't tested their DNA—had a child you don't know about. Responding to this issue will require awareness of your family's dynamics

and a responsible approach to assisting—or possibly not, depending on the situation—the new relative.

By following these easy steps, it is possible to anticipate and be prepared for most of the ethical issues that can arise from genetic genealogy testing, ensuring that DNA testing is as rewarding as possible for everyone.

There is no question that ethical guidelines such as the Genetic Genealogy Standards create real roadblocks to recruitment, testing, and research. However, in order to promote and support a tool that benefits everyone, it is necessary to create these roadblocks for those who don't understand all the potential outcomes of DNA testing.

The Genetic Genealogy Standards cannot anticipate, prevent, or resolve every ethical issue that might be encountered by a genealogist or DNA test-taker. However, the Standards—and an understanding of DNA testing in general—can help educate test-takers about the possible outcomes of DNA testing. Armed with this knowledge, potential test-takers are able to make informed decisions about DNA testing, thereby preventing many of the issues that can arise.

CORE CONCEPTS: ETHICS AND GENETIC GENEALOGY

- **Genetic exceptionalism is the theory that genetic information is unique and must be treated in a different way from other genealogical information. However, all genealogically relevant record types—including DNA—are capable of revealing known and unknown information, including family secrets. As a result, many genealogists reject the idea that DNA should receive special or different treatment.**

- **The only way to prevent the inadvertent disclosure of genealogical information—such as long-forgotten or hidden secrets potentially revealed by DNA testing—is to prevent all genealogical research.**

- **The Genetic Genealogy Standards were created to provide ethical guidance to genealogists and potential test-takers. These Standards help establish best practices for genealogists using DNA testing, and they're designed to ensure all participants in a DNA study have consented and to protect individuals' data and privacy.**

PART TWO
Selecting a Test

4

Mitochondrial-DNA (mtDNA) Testing

Are you frustrated by all those female ancestors with missing maiden names, or not sure which Edith is your great-great-grandmother? Look no further than the answers provided by mitochondrial DNA (mtDNA). One of the most powerful tools available to genetic genealogists, mtDNA offers a glimpse into the maternal lines of even your most challenging ancestors. mtDNA is so powerful that it is used by the military to identify the remains of recovered soldiers, by scientists to identify the remains of kings and tsars, and by genealogists to solve innumerable genealogical mysteries. So how can mtDNA help you?

Mitochondrial DNA

Mitochondria are tiny energy "factories" located inside almost every cell in the body. These factories spend every hour of every day producing energy that the body uses to power things like muscles. You have hundreds or thousands of mitochondria in each cell, and each one contains hundreds of copies of mtDNA. There is a lot of mtDNA in every cell!

Mitochondrial DNA (image A) is a small circular piece of DNA made up of a chain of approximately 16,569 pairs of special molecules called nucleotides. The DNA codes

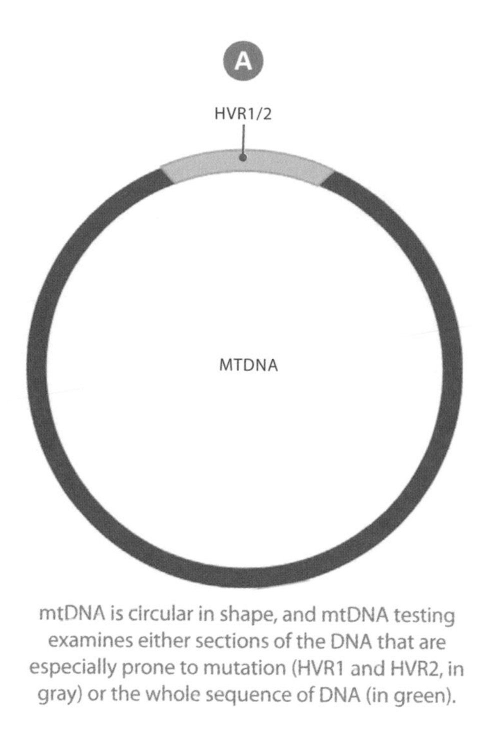

mtDNA is circular in shape, and mtDNA testing examines either sections of the DNA that are especially prone to mutation (HVR1 and HVR2, in gray) or the whole sequence of DNA (in green).

for thirty-seven genes, many of which are directly involved in helping the mitochondria produce energy for the cell. Although mtDNA is an unbroken loop of DNA, scientists and testing companies have given portions of the loop different names based on the DNA found within those portions. The first and second portions, called **hypervariable control region 1 (HVR1)** and **hypervariable control region 2 (HVR2)**, are regions of the mtDNA that accumulate changes relatively quickly, and thus tend to be hyper-variable (i.e., more likely to change) from one person to the next unless those people are closely related. The third portion, the **coding region (CR)**, accumulates far fewer changes and contains the nucleotide base-pair sequence for mitochondrial genes.

The exact start and stop positions for these portions can vary from one testing company to the next, but the most commonly used start and stop positions for each region are:

- HVR1: base pairs 16,001–16,569
- HVR2: base pairs 001–574
- CR: base pairs 575–16,000

As shown in image B, the HVR regions are found on either side of the first numbered base pair (00001) of the mtDNA sequence. There is nothing special about this base pair; it is always counted as the first base pair because it was so identified in the very first mtDNA sequence obtained, and the designation has stuck.

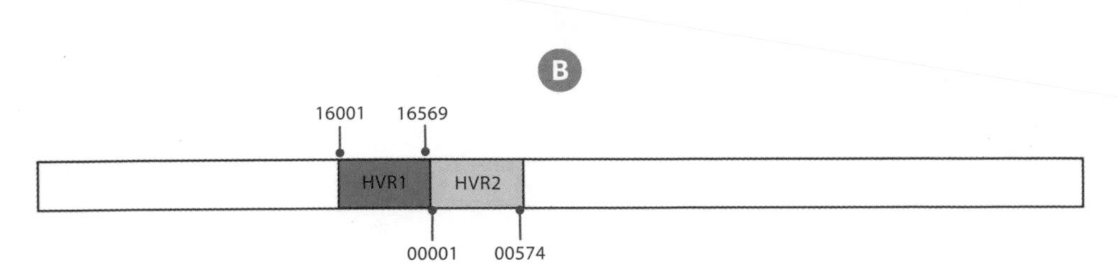

HVR1 and HVR2 are groups of base pairs in mtDNA that are more likely to mutate than the rest of the molecule.

Traditionally, mtDNA testing only sequenced the HVR1 and HVR2 regions. But as the price of sequencing has dropped, most current mtDNA tests sequence all 16,569 base pairs of mtDNA. Full mtDNA sequencing offers several benefits over HVR1/HVR2 sequencing, including better ancient origin information as well as more accurate cousin matching. Reading and comparing the entire mtDNA sequence provides as much information as can be gleaned from this type of DNA. To put it another way, testing the HVR1/HVR2 regions is like reading the abbreviated study guide for *Moby-Dick*, while testing HVR1/HVR2 and the coding region is like reading the entire novel.

The Unique Inheritance of mtDNA

mtDNA has a unique inheritance pattern that makes it particularly valuable for genetic genealogy testing. Unlike other types of DNA, which can be jumbled in a process called **recombination** (more on that later), mtDNA is always passed down from a mother to her children—both male and female—without jumbling. The mother makes exact copies of her mtDNA and passes them down in her egg.

Although mothers pass down mtDNA to both sons and daughters, only daughters will pass it on to the next generation. While every man has mtDNA he inherited from his mother and can be tested, that mtDNA ends with him. He does not pass it on to the next generation.

Image C shows the path of mtDNA inheritance within a short family tree. Joan decides to test her mtDNA and would like to determine from whom in her family tree she inherited that particular piece of DNA. She inherited the mtDNA from her mother, Karen, who in turn inherited it from her mother, Lisa, who in turn inherited it from her mother, Marie. At every generation, only one ancestor carried the mtDNA. And due to this inheritance pattern, Joan will know exactly which ancestor passed down her mtDNA even though she may not know that ancestor's name. For example, Joan has 1,024 ancestors at ten generations (512 men and 512 women), but only one of those 512 women passed down her mtDNA to Joan.

Knowing the inheritance pattern of mtDNA also gives genealogists the ability to trace this piece of DNA forward through a family tree. Joan is a great-grandmother and would like to know which of her descendants carry her mtDNA. Image D is Joan's family tree, in which all the purple-labeled individuals carry Joan's mtDNA. Of course, all four of Joan's children—one son and three daughters—carry her mtDNA. At the grandchild level, four of Joan's five grandchildren carry her mtDNA; her son did not pass it on to the next generation. At the great-grandchild level, only two of Joan's five grandchildren carry her mtDNA, great-grandchildren 3 and 4.

Although Joan's four male descendants in the chart who carry her mtDNA (the four purple square boxes) can take an mtDNA test, none of these males passed on this piece of DNA to the next generation. For mtDNA, a male is a dead end in the line, although they should never be overlooked as a possible testing source. Indeed, a male may be the last living person available to take an mtDNA test for a specific ancestor.

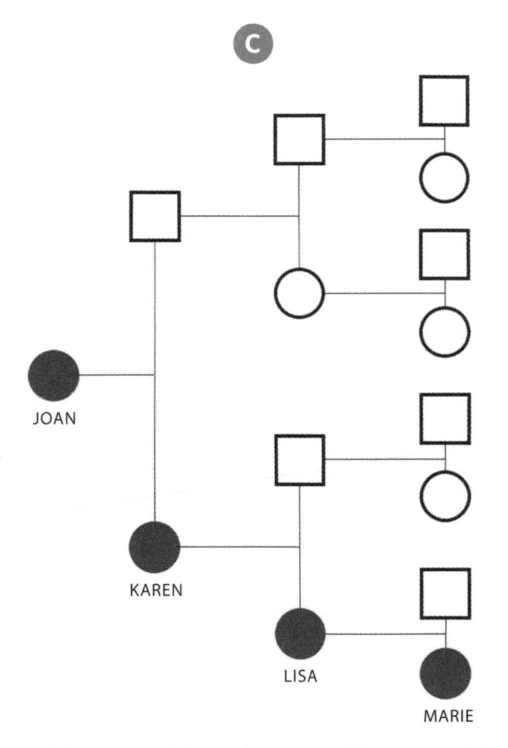

mtDNA is passed down the maternal line (in purple).

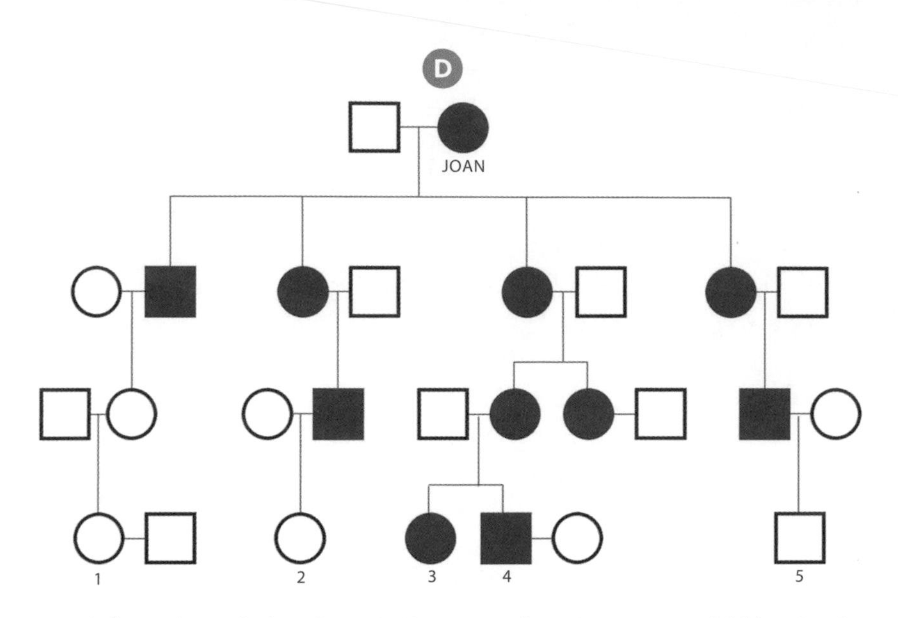

Joan's descendants who have her mtDNA are in purple; notice great-grandchildren 3 and 4 have Joan's mtDNA, but 1, 2, and 5 don't.

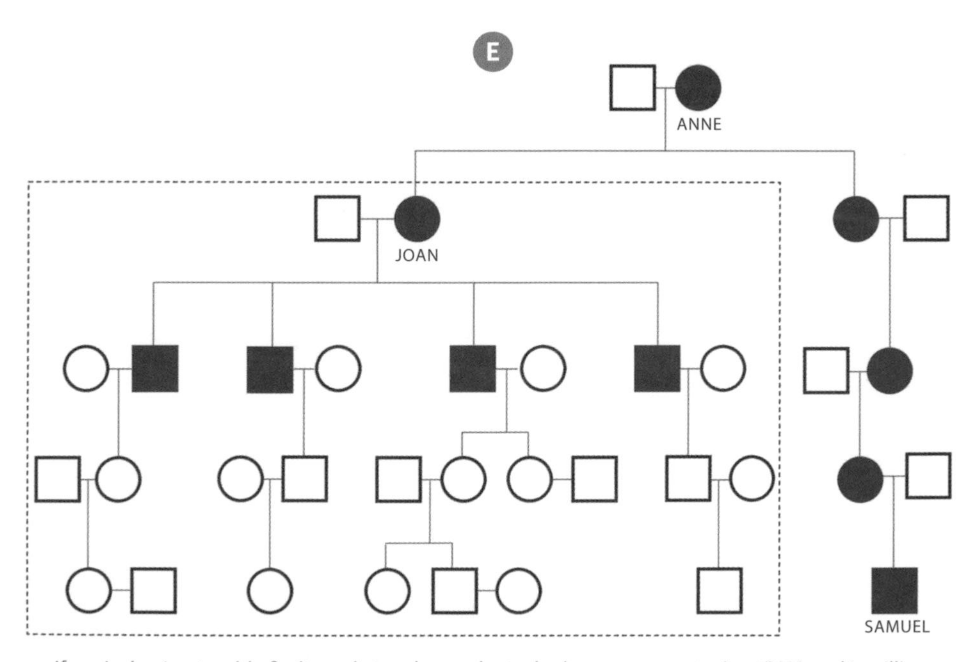

If you're having trouble finding a living descendant who has your ancestor's mtDNA and is willing to take an mtDNA test, work back another generation to find a more distant cousin who can help. Here, Samuel has the same mtDNA as Joan even though he's not one of her direct descendants.

Finding an mtDNA Descendant: Working Backward to Go Forward

To find a living descendant of an ancestor who can take an mtDNA test, a genealogist must trace the mtDNA line through the generations that separate the ancestor and the living descendants. Sometimes, however, an ancestor may have no descendants who carry her mtDNA, even if they have numerous descendants. For example, as shown in image E, Joan had four sons (all deceased), and thus there are no living descendants with Joan's mtDNA and no one who the genealogist can ask to take that test. However, the genealogist may still find a relative who possesses Joan's mtDNA by going back a generation and working forward to determine whether there are any living mtDNA descendants. In this example, Samuel possesses the same mtDNA as his great-great-grandmother Anne and his great-grandaunt Joan. As a result, he can take the mtDNA test to be matched to those two ancestors.

If Samuel is unwilling or unable to take a DNA test, the genealogist will be forced to find another of Anne's descendants or go back yet another generation to Anne's mother's descendants. Sometimes, you may have to go back generations before identifying a suitable mtDNA descendant. There is no limit to how many generations back a genealogist can go to find an mtDNA relative, although the difficulty of researching the maternal line can be a barrier, as the surname usually changes with every generation.

How the Test Works

There are two types of mtDNA tests (image F). The first is **mtDNA sequencing**, which is performed by sequencing all or a portion of the mtDNA genome. A sequenced section of DNA is a long series of just four different letters (*A*, *C*, *G*, and *T*) that stand for the four different nucleotides (adenine, cytosine, guanine, and thymine) that make up all DNA. All mtDNA, for example, is a sequence of 16,569 total base pairs represented by the four nucleotides—*A*, *C*, *G*, and *T*—that make up the base pairs. Low-resolution mtDNA tests sequence just the 1,143 or so base pairs of the HVR1 and HVR2 regions, while high-resolution mtDNA tests sequence every one of the 16,569 base pairs.

The second type of mtDNA test, called **SNP testing**, examines single nucleotide polymorphisms (SNPs) at hundreds or thousands of locations along the circular mtDNA. An **SNP** is a single nucleotide of DNA that can vary from one person to the next. For example, the nucleotide at position 15,833 of the mtDNA may be a cytosine (*C*) in one person or a thymine (*T*) in another person. People who are very closely related should have the same SNP at every location. The more distant the genealogical relationship between two people on the maternal line, the more differences there will be in the tested SNP locations.

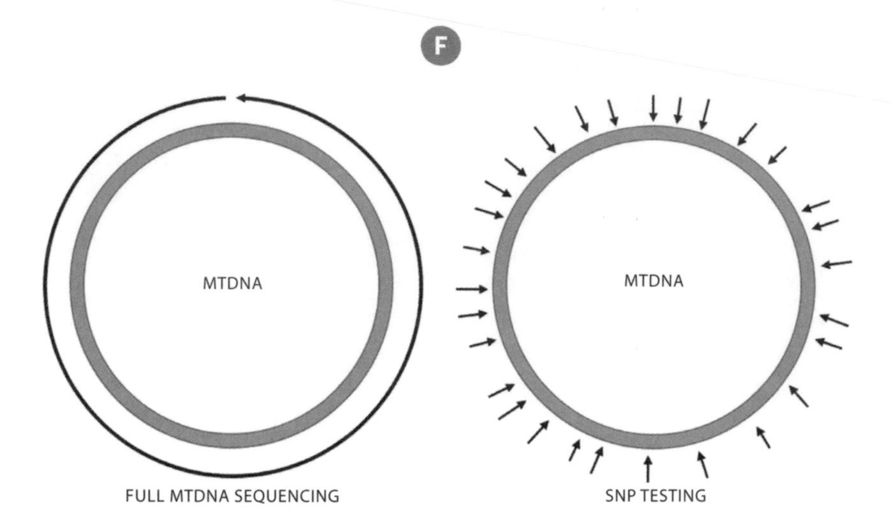

Genealogists have two types of mtDNA to choose from: mtDNA sequencing that looks at the whole mtDNA genome and SNP testing that examines specific portions.

Once the mtDNA is tested by one of these two methods, it is compared to a reference mtDNA sequence, and any differences between the test-taker's mtDNA and the reference mtDNA sequence are identified and listed. Researchers can use three different reference sequences to compare the test-taker's mtDNA:

- The **Cambridge Reference Sequence** (CRS) represents the first mtDNA sequence ever published. This first mtDNA sequence was derived from the placenta of a European female and was published in 1981. It was the only reference mtDNA sequence for several decades.
- The **revised Cambridge Reference Sequence** (rCRS) is an update to the CRS. In the nearly twenty years following the creation of the CRS, researchers discovered several errors, such as missing nucleotides, that were corrected in the rCRS.
- The **Reconstructed Sapiens Reference Sequence** (RSRS) is a recent effort to represent a single ancestral genome of all living humans. The RSRS was introduced in 2012, and scholars are still debating whether to stay with the rCRS or adopt the RSRS. Both sequences have merit and are used in some tests. At Family Tree DNA **<www.familytreedna.com>**, for example, the test-taker's mtDNA is compared to both the rCRS and the RSRS.

Any difference between the test-taker's mtDNA and the selected reference sequence is identified and listed as a **mutation**. Although the word can sometimes have a negative connotation, "mutations" to geneticists are simply changes. The change can involve one nucleotide switching to another, an extra nucleotide appearing, or a nucleotide disappearing, among others. Almost all of these changes are completely benign and harmless, although occasionally a mutation can affect the individual's health, ability to function, or appearance.

The differences between the test-taker's mtDNA and a reference sequence, which are used to determine how closely related two people are on their maternal lines, can be reported in several different ways. For example, differences between the tested mtDNA and the rCRS are acknowledged in the following ways:

- When the mtDNA contains a different nucleotide than the reference sequence, the nucleotide difference is indicated with *the site of the location and the abbreviated nucleotide*, such as *538C* for a cytosine that has replaced the reference nucleotide at position 538. Sometimes the result will provide the reference nucleotide that was replaced, such as *A538C* for a cytosine that replaced the adenine of the reference sequence at position 538.
- When the mtDNA lacks a nucleotide that is present in the reference sequence, the nucleotide difference is indicated by *the position number and a - sign*, such as *522-* for a missing nucleotide at position 522 of the reference sequence.
- When the mtDNA has an extra nucleotide compared to reference sequence, the mutation is indicated by *the position and a .1*. For example, *315.1C* indicates an extra cytosine located after the nucleotide at position 315 of the reference sequence.

Mutation	What It Means
263G	Unlike the reference sequence, the tested mtDNA has a *G* (guanine) at position 263.
A263G	The tested mtDNA replaced the *A* (adenine) at position 263 of the reference sequence with a *G* (guanine).
309.1C	Compared to the reference sequence, the tested mtDNA has an extra *C* (cytosine) after nucleotide 309.
309.2C	Compared to the reference sequence, the tested mtDNA has a second extra *C* (cytosine) after nucleotide 309.
522-	The tested mtDNA is missing the nucleotide found at position 522 in reference sequence.

Once the testing company obtains the list of differences, it can use the information to learn about the test-taker's ancient ancestry and find maternal relatives, which we'll look at in the next section.

Family Tree DNA is the primary mtDNA testing company, having tested the mtDNA of several hundred thousand customers. Family Tree DNA only sequences mtDNA, and although the company used to offer HVR1/HVR2 sequencing, the primary test available today is the full mtDNA sequence. After obtaining the sequencing results, Family Tree DNA compares the sequence to one of the reference sequences and provides the list of differences, or mutations, to the test-taker. In image G, for example, the sequencing results have been compared to the rCRS, and a list of differences has been provided.

Family Tree DNA also uses the RSRS as a reference sequence. Image H lists the differences between the RSRS and the same test-taker from the rCRS example. This shows the importance of knowing which reference sequence a test-taker's mtDNA is being compared to.

23andMe <www.23andme.com> also tests mtDNA, although it uses SNP sequencing instead of HVR1/HVR2 or full mitochondrial genome sequencing. The current version of 23andMe examines approximately three thousand SNPs located all along mtDNA. 23andMe does not provide the list of differences between the test-taker's mtDNA and the reference sequence, although test-takers can review or download their mtDNA information in order to compare it to a reference sequence themselves.

Applying mtDNA Test Results in Genealogical Research

How can an mtDNA test help your research? An mtDNA test has several important uses for genealogists. For example, the results can be used to determine the ancient origins of the mtDNA and to determine whether or not two people are related on their maternal line. The results of the mtDNA test can also be used to estimate the length of time, since the two tested individuals shared a **most recent common ancestor** (or MRCA). In this section, we'll discuss each of these uses in depth.

Determining an mtDNA Haplogroup

Regardless of the type of mtDNA test, results will reveal information about the location of your maternal line thousands of years ago. For example, knowing the origin of your mtDNA, such as whether your maternal line is European, Asian, or Native American, will often provide important clues about the maternal brick wall you've undoubtedly hit in your research.

The results of an mtDNA test are used to determine which **haplogroup** that mtDNA belongs to. An mtDNA haplogroup is a group of maternally related individuals who have

G

Your Results

RSRS Values | rCRS Values

HVR1 DIFFERENCES FROM rCRS				
16187T	16189C	16223T	16290T	16319A
16362C				

HVR2 DIFFERENCES FROM rCRS				
64T	73G	146C	153G	235G
263G	309.1C	309.2C	315.1C	522-
523-	538C	573.1C	573.2C	

CODING REGION DIFFERENCES FROM rCRS				
663G	750G	1438G	1736G	2706G
4029T	4248C	4769G	4824G	7028T
7124G	8027A	8572A	8794T	8860G
11016A	11719A	12007A	12366G	12705T
13681G	14693G	14766T	15326G	

Family Tree DNA provides a list of specific differences between the test-taker's mtDNA and the rCRS.

H

RSRS Values | rCRS Values

Extra Mutations	309.1C 309.2C 315.1C A538c 573.1C 573.2C C4029T G8572A A12366G A13681G A14693G C16519T
Missing Mutations	T16187C C16189T C16111T

HVR1 DIFFERENCES FROM RSRS			
A16129G	G16230A	T16278C	C16290T
C16311T	G16319A	T16362C	C16519T

HVR2 DIFFERENCES FROM RSRS			
C64T	C152T	A153G	C195T
A235G	A247G	309.1C	309.2C
315.1C	A538c	573.1C	573.2C

CODING REGION DIFFERENCES FROM RSRS			
A663G	A769G	A825t	A1018G
A1736G	A2758G	C2885T	T3594C
C4029T	G4104A	T4248C	T4312C
A4824G	A7124G	G7146A	T7256C
A7521G	G8027A	T8468C	G8572A
T8655C	G8701A	C8794T	C9540T
G10398A	T10664C	A10688G	C10810T
C10873T	C10915T	G11016A	A11914G
G12007A	A12366G	G13105A	G13276A
T13506C	T13650C	A13681G	A14693G

The rCRS and the RSRS are not the same, so it's important to note which template your DNA is being compared to.

HETEROPLASMY

Some mitochondrial test results indicate that the test-taker's mtDNA is **heteroplasmic**, and this can create problems for researchers looking to mtDNA for proof of a relationship with someone else.

Heteroplasmy is the presence of more than one mtDNA sequence in a cell or organism. Because human cells have hundreds or thousands of mitochondria, some of the mtDNA in that cell can possess a mutation that the other mtDNA in that cell don't possess. People or individual cells with two or more different mtDNA sequences are heteroplasmic, while people or individual cells with a single mtDNA sequence are **homoplasmic**.

When a heteroplasmic cell divides, the mtDNA will segregate randomly into the two progeny cells. Over time, a heteroplasmic cell can eventually give rise to a homoplasmic cell, though it can take many, many generations for this to happen.

Heteroplasmy can be detected by a commercial genetic genealogy test in the buccal (cheek) cells of the test-taker, where the mtDNA is obtained for the test. However, a test-taker's heteroplasmy may or may not be found in her offspring due to the random segregation of the mitochondria. Additionally, a heteroplasmy that is present in the mother's cheek cells may not be present in the egg that gave rise to the child, and vice versa. Accordingly, the child of a heteroplasmic parent may have mtDNA with one of three different outcomes:

- **Heteroplasmic:** Each egg cell (which develops into a child) has some mitochondria with the heteroplasmic mutation and some mitochondria without the mutation. When the child takes an mtDNA test, both versions of the mtDNA may be detected.
- **Homoplasmic with the mutation:** In this outcome, all the mitochondria in the egg that became the child had the mutation, or (if the child is in fact heteroplasmic) the buccal cells of the child only have a version of the mitochondria with the mutation.
- **Homoplasmic without the mutation:** Even though the parent's buccal cells are heteroplasmic, the child inherited only mitochondria without the mutation. Alternatively, if the child is in fact heteroplasmic, the buccal cells of the child only have mitochondria without the mutation.

A heteroplasmy is written with the original value in the reference sequence, the location of heteroplasmy, and a symbol that indicates what nucleotides are found at that location. For example, a heteroplasmy of *C* or *G* at position 263 would be written as *A263S*. The table below contains the symbols used to represent various combinations of nucleotides for heteroplasmic results.

Heteroplasmy can affect mtDNA matching through Family Tree DNA. For example, if two individuals have identical mitochondrial genomes with the exception of a heteroplasmic mutation in one of them, they may show up as having a genetic distance of one. Thus, if Bill has mutation *16230A* and John has mutation *16230W* (indicating an *A* or *T* at this location), they are not shown as exact matches.

The most famous example of heteroplasmy is Tsar Nicholas II of Russia (1868–1918). Testing his skeletal remains revealed a heteroplasmy of *C* and *T* at position 16169. (If the results had been reported at Family Tree DNA, this heteroplasmy would be *16169Y*.) The heteroplasmy originally confused researchers trying to prove the skeletal remains belonged to the Tsar, since the heteroplasmy was not found in the Tsar's maternal relatives against which the sequence was compared. However, the same heteroplasmy was later identified in the remains of Grand Duke George Alexandrovich (1871–1899), brother of Tsar Nicholas II. The ratio of the heteroplasmic mutation differed in the two brothers, with the Tsar having mostly *C/t* and his brother George having mostly *T/c*. (The capitalized letter represents the predominant result for the tested base pair at position 16169, and the lowercase letter represents the minority result at that position.)

Symbol	Meaning
B	*C* or *G* or *T*
D	*A* or *G* or *T*
H	*A* or *C* or *T*
K	*G* or *T*
M	*A* or *C*
N	*G* or *A* or *T* or *C*
R	*A* or *G*
S	*C* or *G*
U	*U*
V	*A* or *C* or *G*
W	*A* or *T*
X	*G* or *A* or *T* or *C*
Y	*C* or *T*

MITOCHONDRIAL EVE

If every human on earth could trace back his or her maternal line as far as possible, they all would merge on a single person, a woman called "Mitochondrial Eve" who is the mtDNA ancestor of all living humans. She is the most recent common ancestor of all humans on their maternal line. Indeed, all living humans likely have a much more recent atDNA common ancestor, probably on the order of just a few thousand years ago.

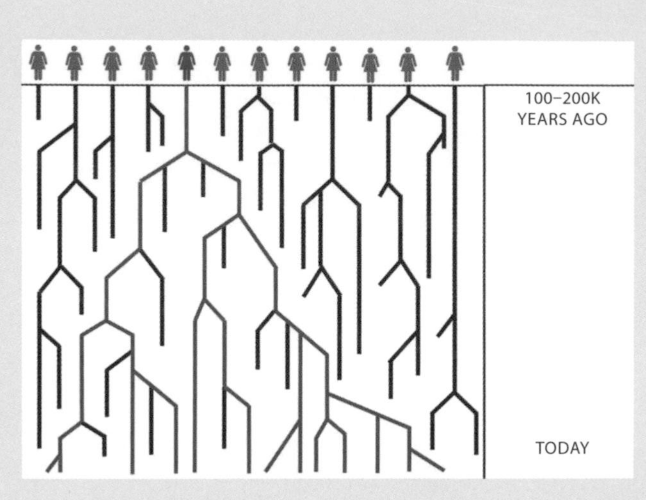

Although we will never know Mitochondrial Eve's real name, we know a few things about her:

1. She probably lived about one to two hundred thousand years ago. The date is based on the variation—the mutations—found in the mtDNA of all of her descendants. Using current information about the mutation rate of mtDNA (approximately one mutation every thirty-five hundred years per nucleotide), it has taken approximately one hundred thousand or two hundred thousand years for all that variation to arise. This can considerably move Mitochondrial Eve's estimated lifetime.
2. She likely lived in East Africa, as the oldest branches of the mtDNA family tree are all found (and appear to have originated) in East Africa.
3. She had at least two daughters, who each gave rise to different lines of the mtDNA family tree. This created a branch point in the mtDNA family tree, as hypothetical Mitochondrial Eve provided one daughter with one type of mtDNA, and a second daughter with a second type of mtDNA.

Although Mitochondrial Eve is named after the biblical Eve, she was not the only woman alive at that time and is not the only one of her contemporaries to have living descendants. It is likely that thousands of other women alive at that time have living descendants, but a final mtDNA descendant in each of these other lines failed to produce daughters at some point between then and today.

New mtDNA lines may be discovered that could further push back the date of Mitochondrial Eve. For example, if a new mtDNA were discovered that pre-dated Mitochondrial Eve based on the number of mutations in the sequence, the date of Mitochondrial Eve would have to be pushed back in time so she could also be the ancestor of the newly discovered line.

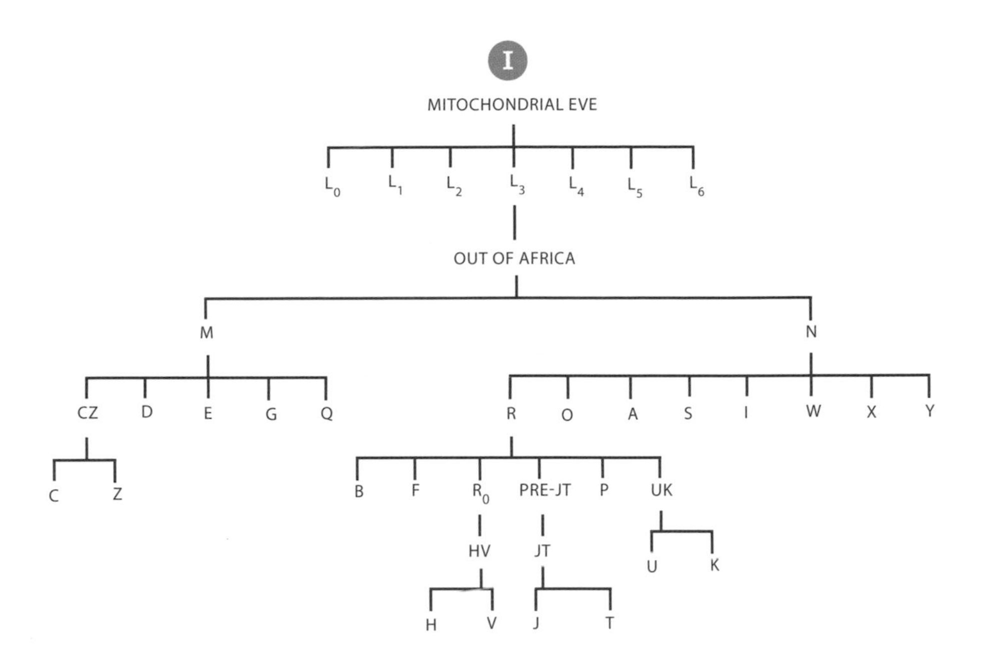

Major groups of mtDNA haplogroups (called subclades) can be mapped in accordance with how they evolved from Mitochondrial Eve.

a recent common ancestor on a particular branch of the mtDNA family tree, which is defined by a particular SNP mutation. (Genealogists can also have a Y-DNA haplogroup, which we'll discuss in chapter 5.) All members of an mtDNA haplogroup can trace their maternal line back to a single ancestor who lived in a specific location several thousand years ago. In most cases, scientists have a good idea of the general location where mtDNA haplogroup ancestors lived.

Haplogroups are named with letters and numbers, and individuals in the same haplogroup will have the same (or a very similar) list of mutations. For example, mtDNA haplogroup *A2w* is a subgroup within haplogroup *A*. The mtDNA haplogroup *A2w* is one of the five mtDNA haplogroups found in North and South American indigenous peoples (the others being *B*, *C*, *D*, and *X*). If someone were to reach a maternal brick wall and learn from an mtDNA test that they belonged in mtDNA haplogroup *A2w*, for example, that test-taker would know to look for a Native American ancestor somewhere along the maternal line, ranging from very recently to long ago.

If all mtDNA sequences on earth were plotted onto a giant family tree, they would all trace back to Mitochondrial Eve (image I; see the Mitochondrial Eve sidebar). From Mitochondrial Eve forward, major branches of the family tree indicate new haplogroups and minor branches indicate subgroups, or subclades, of that new haplogroup. Each branch, whether major or minor, is defined by a particular SNP mutation. Although some SNP mutations are found in multiple branches, usually a branch contains a number of mutations such that an mtDNA sequence can be properly assigned to the correct haplogroup.

Each haplogroup is associated with an approximate time and place in which the founder of that haplogroup arose. This information is based on mutation rates and modern-day distributions of the haplogroup, not on ancient samples of mtDNA, although ancient DNA is being used to further study and refine information about various haplogroups.

Mitochondrial haplogroup *J*, for example, is estimated to have arisen approximately forty-five thousand years ago in the Near East or Caucasus region of the world. In contrast, mitochondrial haplogroup *T* is a newer haplogroup that likely originated approximately seventeen thousand years ago in or around Mesopotamia.

Once test-takers receive their haplogroup assignment, they can seek out more information about that haplogroup and ancient origins. For example, the mtDNA Haplogroups page at WorldFamilies **<www.worldfamilies.net/mtdnahaplogroups>** is a great resource with information about each of the major branches of the mtDNA family tree.

Finding mtDNA Cousins

Another popular use of mtDNA testing is to hunt for mtDNA cousins. At Family Tree DNA, for example, a test-taker's mtDNA is compared to all other mtDNA in the database, and the test-taker will receive a list of anyone in the database who has identical or nearly identical mtDNA. These individuals are mtDNA cousins and are related to the test-taker through the maternal line. Some may have identical mtDNA, while others might differ by one or two mutations. Generally, the fewer the differences between the two sequences, the more closely those two individuals are related.

For example, in image J, six test-takers in the Family Tree DNA database have mtDNA similar to the test-taker's mtDNA. However, all of these individuals have a **genetic distance** of 1 or more, meaning the two mtDNA sequences are not identical; instead, they differ by one or more mutations. In this interface, it is not possible to directly compare your mtDNA with the matches' mtDNA. However, for a genetic distance of 1, for example, either the test-taker's mtDNA has a mutation that the other individual's mtDNA does not have, or it's missing a mutation that the other does have. For example, your mtDNA may be identical to the genetic match's mtDNA, except you have a *T16362C* mutation that the match does not have. Or perhaps the match has a *G16319A* mutation that you don't have. Similarly, with a genetic distance of 2, there are a variety of possible explanations: You might have two mutations that the match does not have, the match may have two mutations that you don't have, or you both may have one mutation that the other does not have.

J

HVR1, HVR2, CODING REGIONS - 6 MATCHES			
Genetic Distance	**Name**		**mtDNA Haplogroup**
1	Riley Gibson	FMS FF	A2w
1	John Johnson	FMS FF	A2w
1	Mary Roberts	FMS FF	A2w
2	Gwen Matthews	FMS FF	A2w
2	Ian Philips	FMS FF	A2w
3	Hiram Culpepper	FMS	A2w

Any mtDNA matches whose DNA has few differences from yours will have a low genetic distance from you, meaning they're likely more closely related to you than are matches with more differences/a higher genetic distance from you.

However, it's difficult to pinpoint how closely related two individuals with mtDNA matches are. Because mtDNA changes relatively slowly, individuals with identical mtDNA can be related either very recently or as much as several thousand years ago. For example, an exact HVR1 and HVR2 match is most likely maternally related somewhere between zero and fifteen hundred years ago. This is one reason why it is better to sequence the entire mtDNA genome rather than just the HVR1/HVR2 regions; an exact full sequence match will likely be maternally related through a common ancestor who lived within the past five hundred years or so.

Type of Match	mtDNA Region Compared at Family Tree DNA	Time to Most Recent Common Ancestor
HVR1 exact match	16,001–16,569 (HVR1)	50-percent chance of common ancestor within about fifty-two generations (1,300 years)
HVR1 & HVR2 exact match	16,001–16,569 (HVR1) and 1–574 (HVR2)	50-percent chance of common ancestor within about twenty-eight generations (700 years)
Full Sequence exact match	16,001–16,569 (HVR1) 1–574 (HVR2) 575–16,000 (Coding Region)	95-percent chance of common ancestor within about twenty-two generations (550 years)

To find the maternal ancestor shared with an mtDNA match, the test-taker can review the match's online family tree or contact the match and ask if he is interested in sharing information. If the match is willing to cooperate, the test-taker can determine whether the two individuals share any names or locations on their maternal lines. Sometimes mtDNA matches will list a match's most distant maternal ancestor, which the test-taker might be able to use to reverse engineer his maternal line if he's not interested in sharing information with others.

In addition to the list of mtDNA matches provided by the company, a test-taker can also search MitoSearch **<www.mitosearch.org>** for others who share their mtDNA. MitoSearch is a free publicly available database with thousands of records from several different testing companies. Test-takers can upload their mtDNA to MitoSearch to look for matches that are already in the database, as well as compare their results to new matches that are uploaded to MitoSearch.

Analyzing Genealogical Questions

In addition to learning about the ancient origins of the maternal line and finding mtDNA relatives, you can use the results of an mtDNA test to help with specific genealogical tasks, such as confirming known lines, analyzing family mysteries, and potentially breaking through brick walls. Traditional documentary research used with the results of mtDNA testing can be a powerful combination for genealogists.

Since mtDNA is inherited maternally, it is very good at determining whether two people are related through their maternal lines. Many genealogical applications of mtDNA, therefore, use mtDNA test results from two or more people to examine whether those test-taker's mtDNA ancestors could have been maternally related.

For example, it is possible to use mtDNA testing to determine whether you might be maternally related to an autosomal-DNA (atDNA) match. As we'll learn later in the book, an atDNA match can be found on any of your ancestral lines, and it is difficult to identify the common ancestor shared with an atDNA match. If an atDNA match also shares your mtDNA, it can significantly narrow down which lines to search for a common ancestor.

As another example, adoptees sometimes use mtDNA testing to assist in their search for their biological family. Finding an exact mtDNA match can potentially point the adoptee toward the biological mother's family, provided the match is closely related to the adoptee and has a well-researched family tree.

Remember both the benefits and limitations of mtDNA testing when applying the results to a genealogical question. For example, mtDNA testing can only determine whether two people are maternally related on their *direct* matrilineal line. Accordingly, an mtDNA test will likely not be the first choice when the genealogical question is whether two men born in the 1800s were brothers. Further, an mtDNA test can only reveal that two people are maternally related somehow, but it can't determine the exact nature of the relationship. As a result, people with matching mtDNA might be sisters, mother/daughter, aunt/niece, first cousins, and so on, for many generations.

These limitations must be contrasted with some of the powerful benefits of mtDNA testing. Unlike atDNA, for example, mtDNA passes down to the next generation unchanged and therefore does not get diluted like atDNA. A test-taker has 100 percent of the mtDNA of her mother's mother's mother's mother (her great-great-grandmother), but just approximately 6.25 percent of her atDNA. Accordingly, even with its limitations, mtDNA can be a powerful tool for genealogists.

CORE CONCEPTS: MITOCHONDRIAL-DNA (MTDNA) TESTING

- **mtDNA is a circular piece of DNA located within the mitochondria of the cell.**

- **Although both men and women inherit mtDNA from their mothers, only women pass down mtDNA to the next generation. As a result of this unique inheritance pattern, mtDNA is only used to examine a test-taker's maternal line.**

- **mtDNA testing is done by either sequencing portions of (or the whole) mtDNA or through SNP analysis of the mtDNA. Full sequencing of the mtDNA is the best test and provides the most information.**

- **The results of any mtDNA test can be used to determine the haplogroup, or ancient origins, of the maternal line back thousands of years.**

- **The results of an mtDNA sequencing test can be used to fish for genetic cousins. However, since mtDNA mutates so slowly, it is not as useful for finding random genetic cousins in a testing company's database. An exact mtDNA match may be very closely related, or may be maternally related hundreds of years ago.**

- **Results from an mtDNA test can be useful for examining specific genealogical questions, such as whether or not two people are maternally related.**

DNA in Action

Are They Sisters?

Say a genealogist has identified three historical women (Mary, Jane, and Prudence) who are potential sisters based on paper-trail evidence. To determine whether the women might have in fact been sisters, the genealogist has traced descendants of Mary, Jane, and Prudence and asked them to take an mtDNA test. All three descendants agreed, and the genealogist is now reviewing the results. Note that any or all of the three descendants could be either men or women, but the line from the three women to their living descendant has to be an unbroken maternal line of mother to daughter.

The (simplified) results in the table show that Mary and Prudence's mtDNA descendants have identical mtDNA but Jane's mtDNA descendant has a genetic distance of 3—that is, there are three differences between the mtDNA of Jane's descendant and the other mtDNA results. As the results show, Jane's descendant is missing two mutations found in the other test-takers, and has one additional mutation not found in the other two test-takers. Since so few generations have passed between the three women and each respective mtDNA descendant, it is unlikely that there was enough time for a genetic distance of 3 to arise.

Mary Smith's mtDNA Descendant	Jane Smith's mtDNA Descendant	Prudence Smith's mtDNA Descendant
73A	73A	73A
146T	146T	146T
315.1C	315.1C	315.1C
16129G	-	16129G
16223C	-	16223C
-	16311T	-

Do the mtDNA test results alone prove that Mary and Prudence were sisters? Unfortunately, the results only establish that they could have been sisters. They also could have been mother/daughter, aunt/niece, maternal cousins, or a variety of other maternal relationships. Indeed, they could even be very distant maternal cousins. The DNA evidence will have to be combined with the traditional documentary evidence in order to create a strong argument that Mary and Prudence were sisters.

Similarly, the results do not definitively prove that Jane could not have been a sister of Mary and Prudence, either. It is possible, although unlikely, that there could have been a misattributed parentage event in the line between Jane and her descendant. For example, there might have been an undocumented adoption in that line. Additionally, it is possible that the mtDNA line between Jane and her descendant could have accumulated the three observed changes. In other words, that line could have both randomly acquired the *16129G* and *16223C* mutations and added the *16311T* mutation, though this is statistically improbable. Additionally, the genealogist may have erred and mistakenly identified an individual for testing who is not an actual descendant of Jane.

DNA in Action

Is This the King? Part I

In 2012, researchers supported by the Richard III Society <**www.richardiii.net**> found a skeleton under a parking lot in Leicester, England. Based on the time frame in which the skeleton was buried, the age of the skeleton upon the person's death (mid-thirties), and physical characteristics including battle wounds and severe scoliosis, the researchers believed that this skeleton could be the remains of King Richard III of England.

Richard III was the final ruler of the Plantagenet dynasty. On August 22, 1485, the thirty-two-year-old Richard was killed at the Battle of Bosworth Field. Richard was buried within the Greyfriars Friary Church in Leicester. However, the location of Richard's grave was ultimately lost through the passage of time.

To determine whether the skeleton was in fact Richard III's remains, researchers wanted to compare mtDNA obtained from the skeleton to mtDNA obtained from Richard's maternal relatives. Genealogists traced descendants of Richard's sister, Anne of York, through seventeen and nineteen generations to identify two living descendants, Michael Ibsen and Wendy Duldig, who took mtDNA tests. The results of Ibsen's and Duldig's full-sequence mtDNA test showed that they have almost identical mtDNA, differing by only a single mutation even though their mtDNA lines diverged nearly five hundred years ago. Their haplogroup is the relatively rare *J1c2c*.

When the results of the skeleton's full mtDNA sequencing were compared to the Ibsen/Duldig results, they were identical with the exception of the single mutation found in Duldig's mtDNA. Together with the other evidence, the researchers definitively concluded that the remains were those of King Richard III. The site of the exhumation is now the King Richard III Visitor Centre where visitors can see the gravesite under glass.

For more about King Richard's DNA testing, see Turi E. King et al., "Identification of the Remains of King Richard III", originally published in *Nature Communications* <**www.nature.com/ncomms/2014/141202/ncomms6631/full/ncomms6631.html**>.

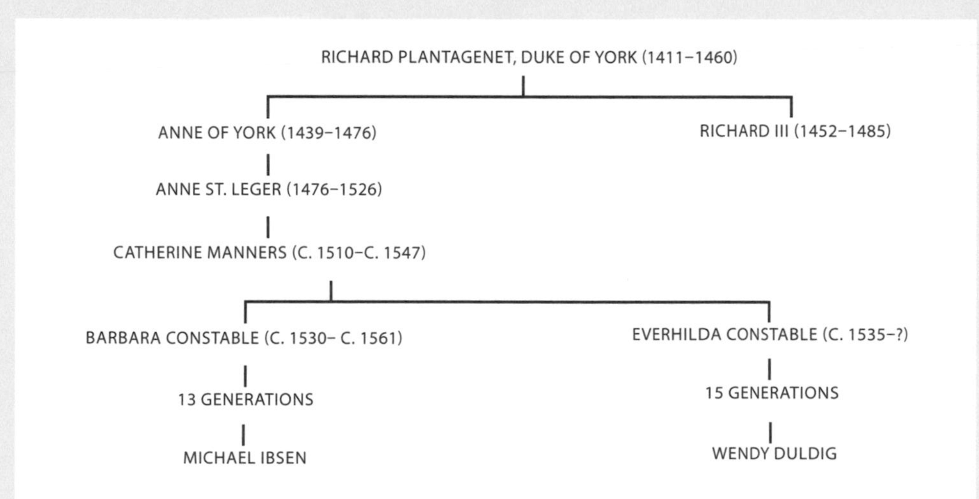

5

Y-Chromosomal (Y-DNA) Testing

Is that Smith ancestor born in Virginia in 1718 the son of John Smith or Hiram Smith? How many men do you have in your family tree who have no identified father, or that after decades of research you are convinced must have been dropped on Earth by aliens? Y-chromosomal (Y-DNA; image A) may be able to help you solve some of these mysteries. Since the Y chromosome is passed down along with the surname in most Western cultures, it is exceptionally useful for examining and breaking through the brick walls of our paternal lines. In this chapter, we'll learn about Y-DNA and how to add this kind of testing to your genealogical toolbox.

The Y Chromosome

The Y chromosome is one of the twenty-three pairs of chromosomes found in the nucleus of a cell and is one of the two sex chromosomes (the other being the X chromosome). While a female has two X chromosomes (see chapter 7 for more about the X chromosome), a male has one X chromosome from his mother and one Y chromosome from his father. As a result, the Y chromosome is found only in men, who inherit it almost entirely unchanged from their fathers.

A

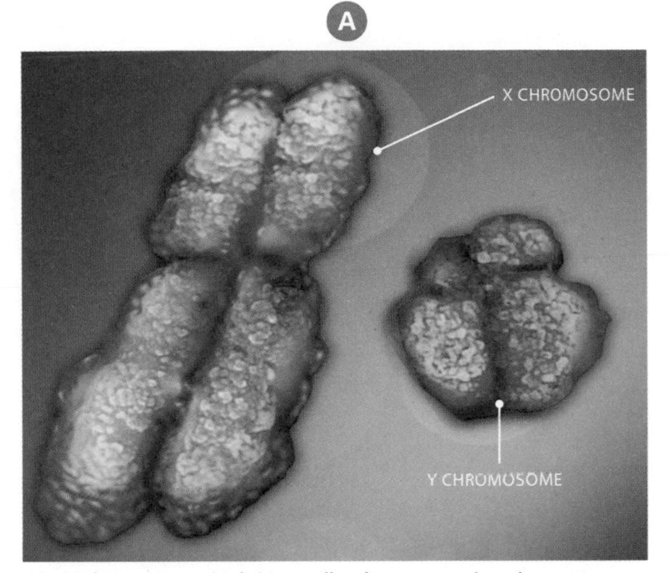

The Y chromosome (right) is smaller than most other chromosomes, including the X chromosome (left), but it contains genetic information valuable to genealogists. Image courtesy of Jonathan Bailey of the National Human Genome Research Institute.

The Y chromosome is approximately fifty-nine million base pairs long, which is actually very short for a chromosome. The chromosome contains approximately two hundred genes, just a small fraction of the estimated twenty to twenty-five thousand genes found throughout the entire human genome.

The Unique Inheritance of Y-DNA

Similar to mitochondrial DNA (mtDNA; chapter 4), Y-DNA has a unique inheritance pattern that makes it valuable for genetic genealogy testing. The Y chromosome is always passed down from a father to his son. The father's cells make an exact copy of this Y chromosome and pass that down to his sons through his sperm. Note that if a man has only daughters, his Y chromosome is not passed on to the next generation.

Unlike all other chromosomes, the Y chromosome is always unpaired, meaning that it does not exchange DNA with another Y chromosome in a process called recombination. Although the tips of the Y chromosome and the X chromosome will sometimes recombine, these regions of the Y chromosome are not utilized for genealogical research or haplogroup determination. As a result, the Y chromosome that a father possesses will almost always be identical to the Y chromosome of his sons.

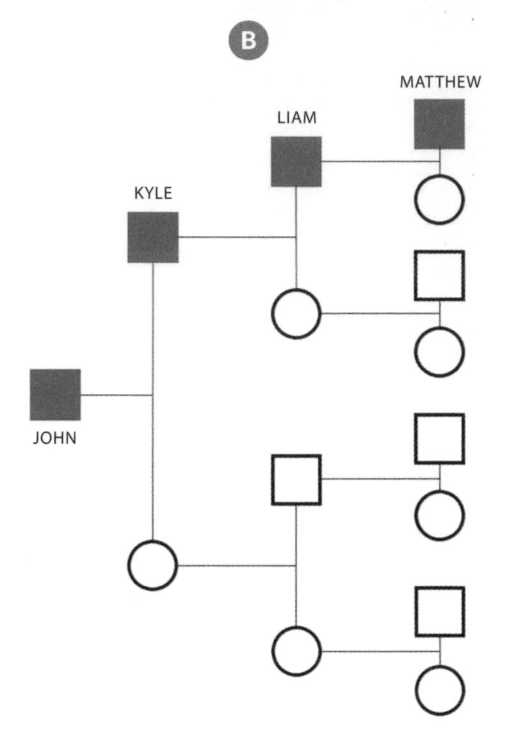

Y-DNA is passed down the paternal line (in blue).

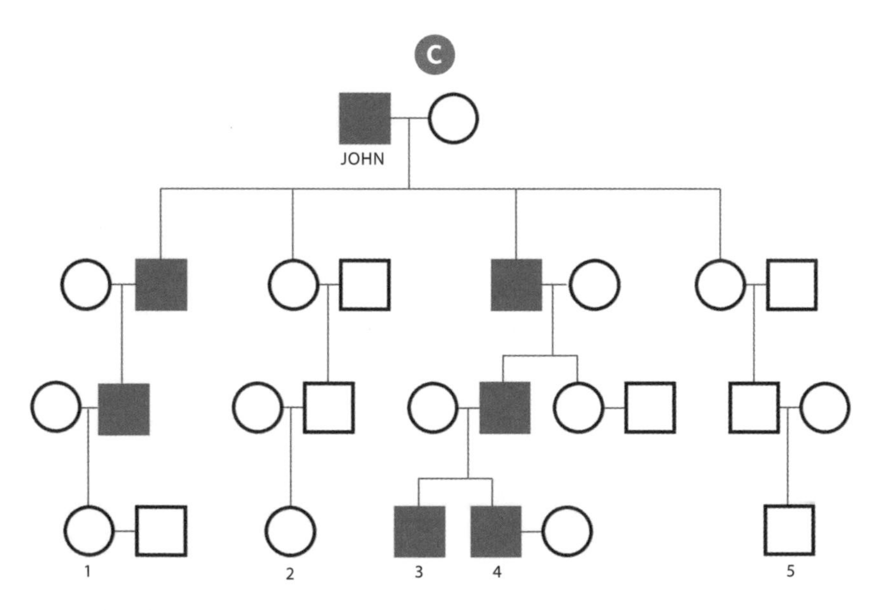

John's descendants who have his Y-DNA are in blue; notice great-grandchildren 3 and 4 have John's Y-DNA, but 1, 2, and 5 don't.

Image B shows the inheritance of the Y-DNA within a family tree. John decides to test his Y-DNA and reviews his family tree to see from whom he inherited that piece of DNA. John inherited the Y-DNA from his father, Kyle, who inherited the same Y-DNA from his father, Liam, who inherited it from his father, Matthew.

At each generation, only one ancestor carried John's Y-DNA, and due to the unique inheritance pattern, John knows exactly which ancestor that is even though he may not know his name or identity. For example, John has 1,024 ancestors at ten generations, a total of 512 men and 512 women. Although every one of those 512 male ancestors had a Y chromosome, only one of them passed down his Y chromosome to John.

Knowing the inheritance pattern of Y-DNA also gives genealogists the ability to trace this piece of DNA forward through a family tree. Let's say John is a great-grandfather and would like to know which of his descendants carry his Y-DNA. Image C is John's family tree, in which blue-labeled individuals carry John's Y-DNA. Only two of John's children, his two sons, carry John's Y-DNA. At the great-grandchild level, two of John's great-grandchildren (3 and 4) carry his Y-DNA. For Y-DNA, a female is always a dead end in the line.

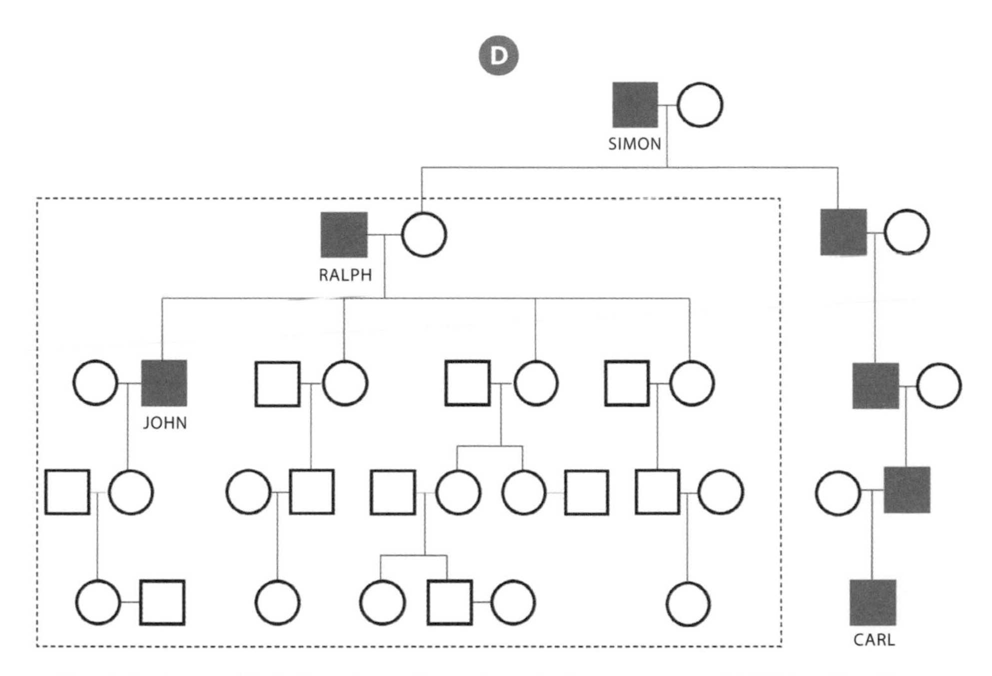

If you're having trouble finding a living descendant who has your ancestor's Y-DNA and is willing to take a Y-DNA test, work back another generation to find a more distant cousin who can help. Here, Carl has the same Y-DNA as John even though he's not one of his direct descendants.

Y-CHROMOSOMAL ADAM

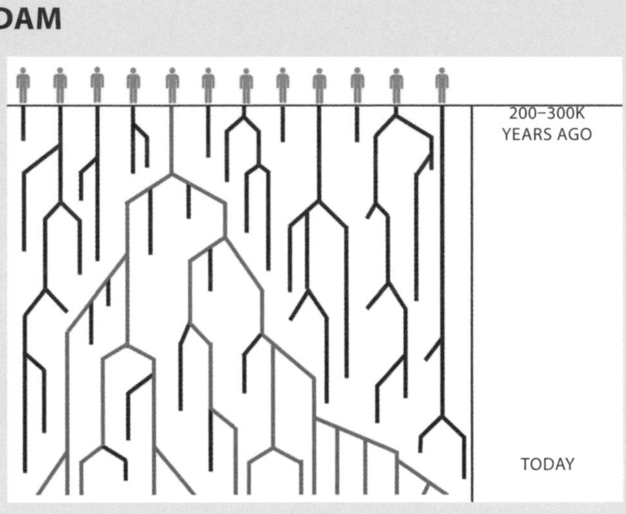

In the previous chapter, we learned that if every human on earth could trace back their maternal line they would all merge on a single person: a woman called "Mitochondrial Eve" who is the mtDNA ancestor of all living humans. Similarly, if every man on earth could trace back his paternal line they would all merge on a single person, a man called "Y-chromosomal Adam." He is the most recent common ancestor (MRCA) of all humans on their paternal line (but not on all lines).

Y-chromosomal Adam's identity has been long since forgotten, yet we know a few things about him. First, we know that he probably lived about two hundred thousand to three hundred thousand years ago based on the number of mutations that are found in modern-day Y chromosomes. Using current information about the mutation rate of Y-DNA, it would have taken approximately two hundred thousand to three hundred thousand years for all the observed variations to arise. Second, we know that Y-chromosomal Adam likely lived in Africa. And third, we know that Y-chromosomal Adam had at least two sons who each gave rise to different lines of the Y-DNA tree.

Similar to Mitochondrial Eve, "Y-chromosomal Adam" has his name's origins in the Bible. However, he was not the only man alive at that time and is not the only one of his contemporaries to have living descendants. Thousands of other men likely lived at that time and have living descendants but at some point between then and today "daughtered out" or had a final Y-DNA descendant in each of these other lines who failed to produce a son.

The date of Y-chromosomal Adam is not a fixed point in time. Even today, old Y-DNA lines are dying out and new Y-DNA lines are being created, which can move the date of Y-chromosomal Adam. Additionally, new Y-DNA lines may be discovered that could push back the date of Y-chromosomal Adam. For example, a paper published in 2012 revealed that an entirely new root haplogroup had been discovered in African-American test-takers, thereby pushing back the date of Y-chromosomal Adam. This new root, called *A00*, is older than the original Y-chromosomal Adam, and therefore a new Y-chromosomal Adam was identified. As more men are tested around the world, particularly in Africa, it is possible that other root haplogroups could be identified and the date of Y-chromosomal Adam could be pushed back further in time.

Daughtering Out

A male can take a Y-DNA test to examine his own Y-DNA line. A female, however, will have to ask her brother, father, or uncle (or another male relative) to take a Y-DNA test. And any genealogist tracing another piece of Y-DNA will have to find a living male descendant who is willing to take a Y-DNA test. Sometimes, however, an ancestor may have no descendants who carry his Y-DNA, even if they have many descendants. In this case, the Y-DNA is said to have "daughtered out."

In the example in image D, Ralph has no living descendants with his Y chromosome. Ralph had one son and three daughters, and his son had only a daughter. Accordingly, Ralph's Y-DNA has daughtered out.

However, it may still be possible to find a relative who possesses Ralph's Y-DNA. By going back a generation and working forward to determine whether there are any living Y-DNA descendants, a genealogist may find a living male relative willing to take a Y-DNA test. In this example, Ralph's father, Simon, possessed the same Y-DNA as Ralph and passed it down to Ralph's brother and down through a line to the living male descendant, Carl.

If Simon had no male descendants who carried his Y-DNA or no descendants willing to take a DNA test, a genealogist could go back yet another generation and work forward. There is no limit to how many generations back a genealogist can go to find a Y-DNA relative, although, as discussed later, a genealogist should consider the increasing possibility of misattributed parentage with every additional generation.

How the Test Works

Normally, the Y chromosome is transmitted from one generation to the next almost entirely without change. Over time, however, the Y chromosome can accumulate one or more mutations that—while typically harmless and not affecting a man's health—can be detected by a test and be useful for genealogical analysis.

Two Y-DNA tests for genealogy are available: **Y-STR tests** and **Y-SNP tests** (image E). Y-STR tests, or "Short Tandem Repeat" tests, sequence between 12 and 111 (and sometimes even more) very short segments of Y-DNA at locations all along the Y chromosome. Similarly, the Y-SNP test, or "Single Nucleotide Polymorphism" test, examines between one and hundreds of single spots along the Y chromosome. In this section, we'll discuss how these tests work and the pros and cons of each.

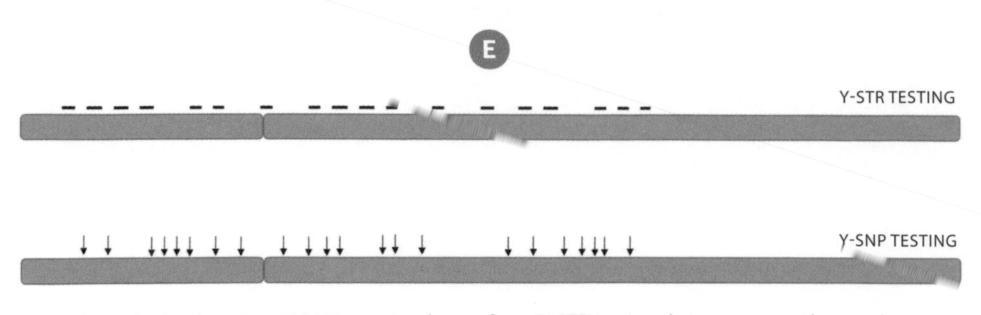

Genealogists have two Y-DNA tests to choose from: Y-STR testing that sequences the number of repeated segments and Y-SNP testing that tests specific points of DNA (SNPs).

Y-STR Testing

Y-STR markers, central to this type of Y-DNA testing, are identified by their DNA Y-chromosome Segment (DYS) number and measured by the number of repeats of a particular DNA sequence at a particular location. The results of Y-STR testing are usually presented with a DYS name and the number of repeats for that particular marker.

The DYS name identifies which specific location along the Y chromosome is being analyzed, and the number of repeats identifies how many repetitions of a nucleotide sequence are found at the location being analyzed. For example, DYS393 is an STR located at a specific position on the Y chromosome, and usually has between nine and eighteen repeats of the sequence *AGAT*, with thirteen repeats being the most common. A DYS393 result of 9, for example, means that there are nine repeats of the *AGAT* sequence at that location:

...ATAC**AGATAGATAGATAGATAGATAGATAGATAGATAGAT**ACTA...
1 2 3 4 5 6 7 8 9

The results of multiple Y-STR markers are typically presented in a table with the DYS marker name in the top row and the number of repeats for each marker in the next row.

DYS#	393	390	19	391	385	426	388	439	389I	392	389II
Repeats	14	23	15	11	11-15	11	13	12	13	13	29
Estimated haplogroup is *R1b1b*											

Together, the results of an individual's tested Y-STR markers represent the individual's **haplotype**, the collection of specific marker results that characterize that test-taker. Every male has a specific Y-DNA haplotype, and generally the more similar the haplotypes of two males, the more closely related they are.

Most Y-STR tests examine between 37 and 111 STR markers, but many more STRs are being identified and used for testing. Family Tree DNA currently offers 37-marker, 67-marker, and 111-marker Y-STR tests. The 67-marker test, for example, contains all of the markers from the 37-marker test, plus an additional thirty markers. Similarly, the 111-marker test contains all of the markers from the 67-marker test, plus an additional forty-four markers. The more Y-STR markers that are tested, the greater the resolution of the estimated relationship between two compared males.

The number of repeats at a particular Y-STR can change over time at a relatively regular rate, thereby giving genealogists the ability to trace patrilineal lineages over time. A father and son, for example, will almost always have the same Y-DNA haplotype. Occasionally, a mutation occurs in one or more of the Y-STR markers between one generation and the next. For example, nine repeats at DYS393 can become ten or even eleven repeats due to a random error. The rate of errors is relatively regular, meaning that the differences between two haplotypes work as a "clock" to estimate how many generations have passed since two men had a common ancestor. DYS393 has a very slow mutation rate of 0.00076, or approximately one mutation in every 1,315 transmission events, on average. However, despite this slow mutation rate, a mutation in DYS393 can randomly occur at any time, leading to a father and son differing at this marker.

Some Y-STR markers have a tendency to change more rapidly than others. In contrast to DYS393's slow mutation rate, for example, DYS439 has a mutation rate of 0.00477, or about one mutation in every 210 transmission events, on average. When comparing the Y-STR results of two men, consider whether they differ at fast markers and/or slow markers. For example, if the two men differ at only "fast" markers, it is likely—but not guaranteed—that their common ancestor could be significantly more recent than two men who differ at only "slow" markers. Family Tree DNA identifies "faster changing STR markers" in Y-DNA surname projects by highlighting them in red. See **<www.familytreedna.com/learn/project-administration/gap-reference/colors-y-dna-results-chart-heading>** for more information.

The sample results also identified the Y-DNA haplogroup of the test-taker as *R1b1b*, though this is just an estimate based on the results of the Y-STR test. (Similar to mtDNA haplogroups, Y-DNA haplogroups are named by letters of the alphabet, and a particular Y-DNA haplogroup result can provide information about the ancient origins of the test-taker's patrilineal line. However, haplogroups can only be estimated by Y-STR testing and are actually defined by Y-SNPs.)

And what can these tests do for you? Y-STR tests are essential for estimating the relatedness between two males. Since Y-STRs exhibit a relatively constant mutation rate, the number of differences between the Y-STR profile of two people—their haplotypes—can be used to estimate the time since those two people shared a common male ancestor. One mutation will mean a more recent common male ancestor, while ten mutations will mean a very distant common male ancestor. Accordingly, Y-STR results are extremely useful for examining genealogical questions involving male lines.

Y-SNP Testing

Y-SNP testing examines hundreds or thousands of SNPs—variable nucleotides *A*, *T*, *C*, and *G*—all along the length of the Y chromosome. Y-SNPs are traditionally used to determine a test-taker's Y-DNA haplogroup and ancient ancestry, but not as useful for finding genetic cousins in the testing company's database. However, new tests are identifying SNPs that may be useful on a genealogically relevant time frame. These so-called "family SNPs" are mutations that developed within the past few hundred years. While there is no test available specifically for family SNPs at the time of this book's publication, these types of tests will probably be available in the near future.

The results of a Y-SNP test can have several important uses. For example, Y-SNP testing accurately determines the test-taker's Y-DNA haplogroup and reveals information about the ancient ancestry of the patrilineal line. Since SNPs are used to define Y-DNA haplogroups, the results of a Y-SNP test can also confirm an estimate or redefine a haplogroup estimate that is based solely on Y-STR results.

In addition, each SNP in the results helps place the test-taker on a branch of the human Y-DNA tree. Every SNP result will be either **ancestral**, meaning the test-taker does not have a mutation at the particular SNP, or **derived**, meaning the test-taker is mutated at that SNP. SNPs and their ancestral or derived classification help define the test-taker's location on the human Y-DNA haplogroup tree. For example, an individual will be derived for the SNPs that define the branch of the Y-DNA haplogroup tree where they belong.

In the following table, for example, the test-taker's Y-SNP test results reveal that his Y-DNA belongs to haplogroup *R1b1a2a1a1*, one of the most common Y-DNA haplogroups in Europe. In this example, the first SNP result, *M269+*, indicates that the test-taker is derived at that SNP. At L277, however, the test-taker is ancestral (hence, *L277-*).

RESEARCH TIP

Note the Names

In the past, the naming convention for Y-DNA haplogroups added a number or a letter for each new branch of the tree. However, as new branches were discovered, the haplogroup names became too long to be useful. A new naming convention, called the terminal SNP, is now most often used for haplogroup naming. For example, the test-taker's terminal SNP in the *R1b1a2a1a1* example is *R-U106*, which is the most distant branch to which he can be mapped.

Haplogroup	SNP Results	Terminal SNP
R1b1a2a1a1	M269+ L23+ L151+ U106+ L277-	R-U106

In this simplified Y-DNA haplogroup tree, the most distant branch of the tree to which the test-taker can be mapped is *R-U106*, otherwise known as *R1b1a2a1a1* (image F).

The human Y-DNA haplogroup tree is far from complete. New branches are constantly being discovered as more men undergo Y-DNA testing. Returning to the previous example, if a new branch of the Y-DNA tree were to be discovered underneath U106, and the test-taker was derived at the SNP that defined that new branch, his terminal SNP would change to the more distant branch of the tree (*R-NEWSNP*; see image G).

The International Society of Genetic Genealogy (ISOGG) maintains an extensive Y-SNP Index **<www.isogg.org/tree/ISOGG_YDNA_SNP_Index.html>**, as well as a detailed Y-DNA Haplogroup tree **<www.isogg.org/tree/index.html>** with a separate page for each

F

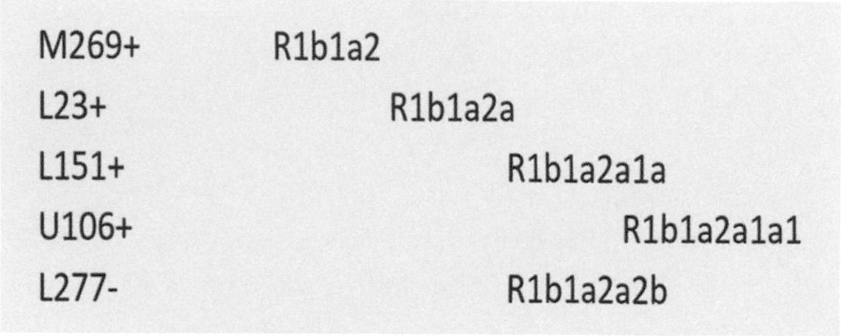

M269+ R1b1a2
L23+ R1b1a2a
L151+ R1b1a2a1a
U106+ R1b1a2a1a1
L277- R1b1a2a2b

Y-DNA haplogroups (such as these that start with the letter *R*) are assigned based on whether an individual is ancestral or derived at various DNA segments (SNPs).

G

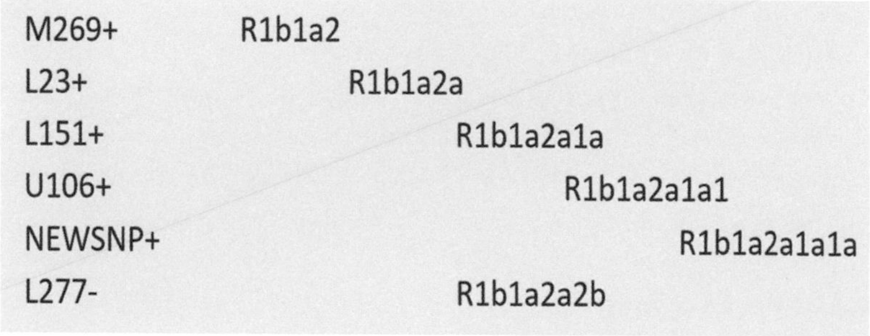

M269+ R1b1a2
L23+ R1b1a2a
L151+ R1b1a2a1a
U106+ R1b1a2a1a1
NEWSNP+ R1b1a2a1a1a
L277- R1b1a2a2b

New branches (such as *R1b1a2a1a1a)* are constantly being added to the Y-DNA haplogroup tree.

haplogroup. In addition to a map of the tree for each haplogroup, the ISOGG site includes a brief description of the haplogroup's origin, a list of primary references, and a list of additional resources.

Applying Y-DNA Test Results to Genealogical Research

Y-DNA tests have many important applications to your genealogical research. For example, the results of a Y-DNA test can be used to determine the Y-DNA haplogroup of a particular line, find DNA cousins or paternal ancestors, and answer genealogical questions. Y-DNA can also estimate the length of time since two men shared an MRCA on their direct patrilineal line and be used to determine, for example, whether two men could have been brothers or father/son, among other relationships. In this section, we'll discuss each of these uses in depth.

Determining a Y-DNA Haplogroup

The results of Y-STR testing will provide a haplogroup estimate, while the results of Y-SNP testing will provide a more definitive haplogroup determination. All Y-DNA haplogroups, which are named with letters and numbers, descend from Y-chromosomal Adam. From Y-chromosomal Adam forward, major branches of the Y-DNA family tree indicate new haplogroups and minor branches indicate subgroups (or **subclades**) of that new haplogroup (image H). Each branch, whether major or minor, is defined by one or more SNP mutations. Although some SNP mutations are found in multiple branches, usually a branch contains a number of mutations so a Y-DNA sequence can be properly assigned to the correct haplogroup.

The test-taker can then utilize the haplogroup designation to learn about the ancient origins of the direct patrilineal line. For example, the Y-DNA Haplogroups page at World-Families **<www.worldfamilies.net/yhaplogroups>** is a good resource, with brief introductory information about the different Y-DNA haplogroups. Some Y-DNA haplogroups have multiple sources of information available. Occasionally, these sources will have information that appears to be conflicting, but this shouldn't be cause for alarm as researchers are still learning about the human Y-DNA tree. Y-DNA haplogroup descriptions will continue to change as scientists learn more about the Y-DNA tree.

Finding Y-DNA Cousins

You can use the results of a Y-STR test to find genetic cousins who share a direct patrilineal ancestor. The test-taker's Y-STR haplotype—the collection of numerical results at

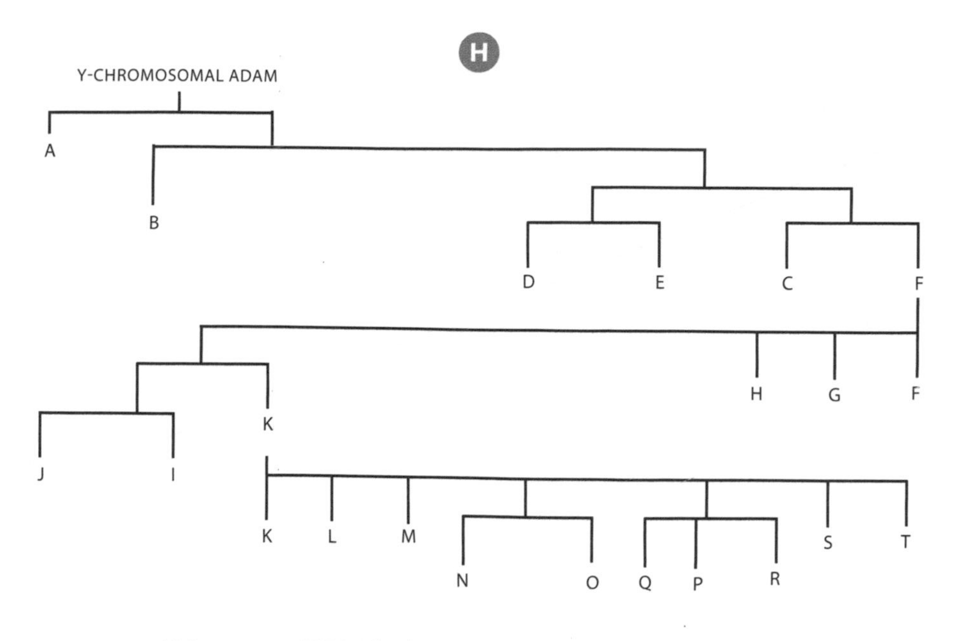

Major groups of Y-DNA haplogroups (called subclades) can be mapped in accordance with how they evolved from Y-chromosomal Adam.

each of the tested markers—is compared to every other Y-STR haplotype in the database, and other test-takers with sufficiently similar results are identified. Usually, two haplotypes must be as close as, or closer than, a minimum threshold set by the testing company in order to be identified in the test-taker's genetic cousin list. The more similar the test-taker's haplotype and the patrilineal cousin's haplotype, the closer the common patrilineal ancestor was to them in time and generational distance.

Only Family Tree DNA offers the ability to compare a test-taker's Y-STR test results to a large Y-STR database. With more than five hundred thousand Y-STR test-takers in the company's database, you're more and more likely to find a patrilineal cousin when taking a Y-STR test.

An individual taking a Y-STR test from Family Tree DNA receives a list of people in the database who have identical or very similar Y-DNA. These individuals are Y-DNA cousins and are related to the test-taker through the patrilineal line. Some may have identical Y-DNA, while others might differ by a handful of STR differences. Generally, the more similar the Y-STR profiles of two men, the more closely those two men are related.

I

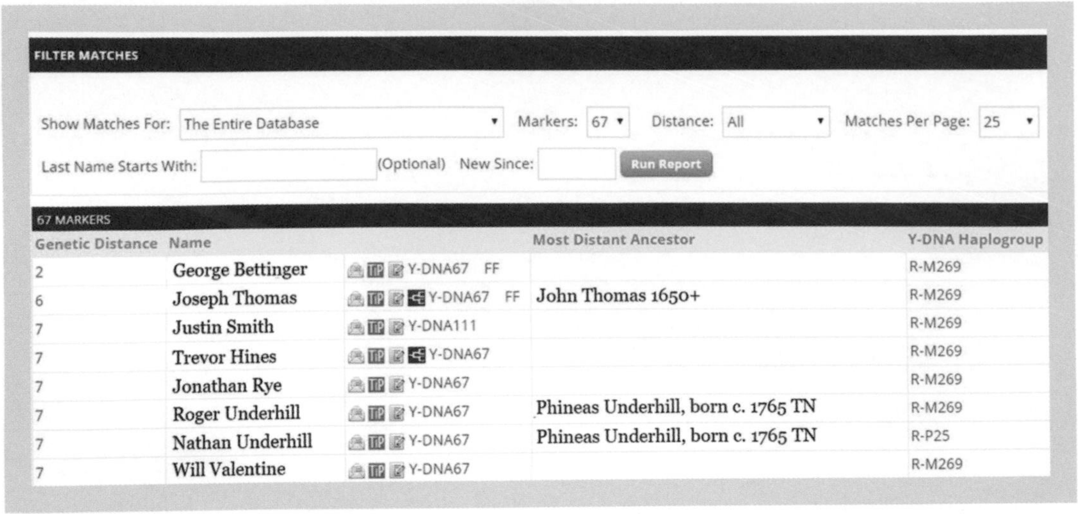

FILTER MATCHES

Show Matches For: The Entire Database | Markers: 67 | Distance: All | Matches Per Page: 25

Last Name Starts With: (Optional) New Since: Run Report

67 MARKERS

Genetic Distance	Name		Most Distant Ancestor	Y-DNA Haplogroup
2	George Bettinger	Y-DNA67 FF		R-M269
6	Joseph Thomas	Y-DNA67 FF	John Thomas 1650+	R-M269
7	Justin Smith	Y-DNA111		R-M269
7	Trevor Hines	Y-DNA67		R-M269
7	Jonathan Rye	Y-DNA67		R-M269
7	Roger Underhill	Y-DNA67	Phineas Underhill, born c. 1765 TN	R-M269
7	Nathan Underhill	Y-DNA67	Phineas Underhill, born c. 1765 TN	R-P25
7	Will Valentine	Y-DNA67		R-M269

These results from Family Tree DNA describe the relationship between the test-taker and his DNA matches, including the genetic distance, haplogroup, and (for some matches) the most distant ancestor.

For example, in image I, the individual has taken a 67-marker Y-STR test from Family Tree DNA. There are eight other test-takers in Family Tree DNA's database who have Y-DNA similar enough to the test-taker's Y-DNA to be shown in the list. However, these individuals have a genetic distance of 2 or more, meaning that the Y-DNA results, or haplotypes, are not identical; instead, they differ by two or more mutations.

Genetic distance is calculated by adding together the difference between the results for each marker where the two test-takers differ. In the following example, the two test-takers differ by a value of 1 at two different markers and have a genetic distance of 2:

Name	DYS#	393	390	19	391	385a	385b	426	388	439
Thaddeus Alden	Results	14	22	15	11	11	15	11	13	9
Thomas Alden	Results	14	23	15	11	11	15	11	12	9

In this example, the two test-takers differ by a value of two at one marker, and thus also have a genetic distance of 2:

Name	DYS#	393	390	19	391	385a	385b	426	388	439
Thaddeus Alden	Results	14	23	15	11	11	15	11	13	9
Thomas Alden	Results	14	23	15	11	11	17	11	13	9

The genetic distance provides insight into the amount of time and number of generations that have elapsed since two test-takers shared a common patrilineal ancestor. For example, at sixty-seven Y-STR markers, a genetic distance of 0 indicates a more recent common ancestor while a genetic distance of 7 indicates a much, much older common ancestor.

The table below (adapted from Family Tree DNA's "Expected Relationships With Y-DNA STR Matches" **<www.familytreedna.com/learn/y-dna-testing/y-str/expected-relationship-match>**). breaks down how closely related Y-DNA matches are across multiple tests, by genetic distance:

	37 Y-STR Markers	67 Y-STR Markers	111 Y-STR Markers	Interpretation
	Genetic Distance			
Very Tightly Related	0	0	0	The relatedness between the two test-takers is extremely close, and few people find or test a cousin at this genetic distance.
Tightly Related	1	1–2	1–2	The relatedness between the two test-takers is very close, and few people find or test a cousin at this genetic distance.
Related	2–3	3–4	3–5	The relatedness between the two test-takers is within the range of most well-established surname lineages in Western Europe, but finding a common ancestor might be challenging.
More Distantly Related	4	5–6	6–7	Without additional evidence, it is unlikely that the two test-takers share a common ancestor within a genealogically relevant time frame.

If a test-taker has only taken a 37-marker test (or an even older test with fewer markers), upgrading to a 67- or 111-marker test could provide additional insight into the genealogical relationship between two men. For example, it is possible that a genetic distance of 2 at thirty-seven markers will remain 2 when upgraded to sixty-seven markers, which suggests a much *closer* genealogical relationship than was observed at thirty-seven markers. In contrast, it is also possible that a genetic distance of 2 at thirty-seven markers will increase to 3 or more when upgraded to sixty-seven markers, which suggests a more distant genealogical relationship than was observed at thirty-seven markers. Genetic distance can increase, but should never decrease, with additional Y-STR testing.

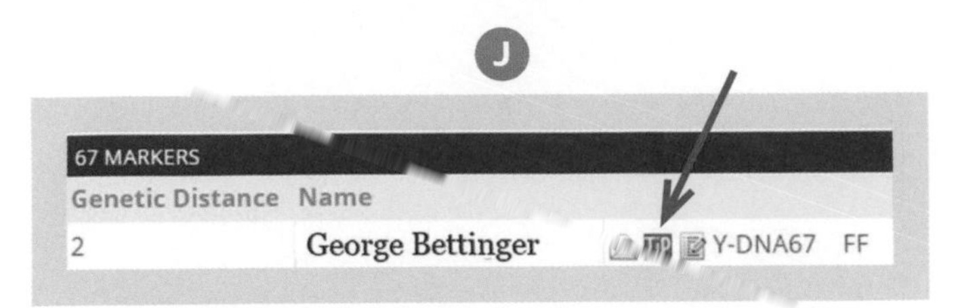

67 MARKERS
Genetic Distance Name
2 George Bettinger Y-DNA67 FF

Family Tree DNA's FTDNATiP will calculate the probability that you shared an ancestor with your match.

In comparing Y-DNA 67 marker results, the probability that Joseph U. Alden and Allen W. Alden II shared a common ancestor within the last...

COMPARISON CHART

Generations	Percentage
4	44.43%
8	84.11%
12	96.58%
16	99.37%
20	99.89%
24	99.98%

FTDNATiP provides percentage estimates that you and a Y-DNA match shared an ancestor within a certain number of generations.

Family Tree DNA also provides a statistical analysis of the distance between two test-takers that have similar Y-STR haplotypes. This statistical analysis is called the Family Tree DNA Time Predictor (FTDNATiP). The FTDNATiP analysis can be performed on any of a test-taker's list of individuals with similar Y-DNA, and can be found by clicking the orange box labeled TiP on the Matches page (image J).

FTDNATiP compares the results of the test-taker and his identified match, and calculates the time to the most recent common ancestor (TMRCA) using a patented algorithm that utilizes the specific mutation rate for each of the markers where there is a difference between the two men.

In the following example (image K), the two test-takers have a genetic distance of 2 at sixty-seven markers. The TiP calculator calculates a 44.43-percent probability that the test-takers share a common ancestor in the last four generations, and an 84.11-percent probability that the test-takers share a common ancestor in the last eight generations. In fact, the two test-takers in this example share a common ancestor at six generations.

Because the TiP calculation is based on individual mutation rates for specific markers, the same genetic distance can have slightly different TiP estimates.

A test-taker can also search a free public Y-STR database called Ysearch <www.ysearch.org>. The site was created and is maintained by Family Tree DNA, and it contains thousands of records from test-takers who tested in the past at companies other than Family Tree DNA. As such, Ysearch offers another opportunity for Y-STR test-takers to find men with similar Y-STR haplotypes.

Joining a Surname or Geographic Y-DNA Project

A Y-DNA project is a collaborative effort to answer genealogical questions using the results of Y-DNA testing. A surname project, for example, brings together individuals with the same (or similar) surname(s), while a geographic project gathers individuals by location rather than by family or surname. Other projects bring individuals together based upon their haplogroup designation. Administrators who are responsible for organizing results, sharing information, and recruiting new members to the group run these DNA groups.

Family Tree DNA hosts more than eight thousand different DNA projects, including both mtDNA and Y-DNA projects. The Williams DNA Project <www.familytreedna.com/groups/williams-dna>, for example, has more than thirteen hundred members. Other projects may only have a few test-takers.

Finding a DNA project is usually very simple. Here are four places to begin your search:

- Family Tree DNA <www.familytreedna.com> provides a box to search by surname or by location or country. Alternatively, projects can be found using the alphabetical listing.
- World Families <www.worldfamilies.net/surnames> hosts numerous surname projects and has a search form that skims through the entire site. World Families also has a list of DNA projects with fifty or more members <www.worldfamilies.net/content/surname-projects-50-members>.
- Cyndi's List provides a partial listing of Surname DNA Studies & Projects <www.cyndislist.com/surn-dna.htm>.
- Search engines are one of the easiest ways to find a surname project. Searching *[SURNAME] DNA Project* will typically identify relevant projects in the search results.

DNA projects can potentially accomplish a number of goals for participants, including:

- estimate relationships between individuals in the project
- confirm or reject the relationship of surname variants

- explore the surname's country or countries of origin
- learn more about the migration of the surname over time
- join a community of other genealogists with similar goals

In addition to these benefits, you'll have a financial incentive to joining a surname or geographic project even before ordering a Y-DNA test. Family Tree DNA offers a testing discount to every member of a DNA project.

Analyzing Genealogical Questions

Similar to mtDNA results, the results of a Y-DNA test can be used to examine genealogical questions, including confirming known lines, analyzing family mysteries, and potentially breaking through brick walls. Traditional documentary research can combine with the results of Y-DNA testing to make a powerful tool for genealogists.

Since Y-DNA is inherited paternally, it is very good at determining whether two people are related through their paternal lines. And unlike mtDNA, Y-DNA can estimate approximately how much time has passed since two people shared a common patrilineal ancestor. And unlike autosomal DNA (atDNA), Y-DNA is passed down to the next generation largely unchanged and does not recombine with other DNA. The Y chromosome analyzed in a living male is virtually identical to the Y chromosome in his paternal great-great-great-grandfather.

While providing numerous benefits, Y-DNA testing also has several important limitations when it is being applied to genealogical questions. For example, Y-DNA testing can only determine whether two people are paternally related on their direct patrilineal line. Further, a Y-DNA test can only reveal whether two men are paternally related *somehow*, but is unable to determine exactly how those two men are paternally related. For example, the men could have been brothers, father/son, first cousins, or a more distant relationship like fifth cousins.

It is also possible to use Y-DNA testing to determine whether you might be paternally related to an atDNA match. As we'll learn later in the book, an atDNA match can be found on any of your ancestral lines, but it is difficult to identify the common ancestor shared with an atDNA match. If that atDNA match also shares your Y-DNA (or the Y-DNA of a paternal relative), then you can significantly narrow down which lines to search for a common ancestor.

As another example, adoptees often use Y-DNA testing to assist them in their search for their biological family. Finding a close Y-DNA match can potentially point the adoptee toward the biological father's family, and can even provide a possible biological surname, which can be an enormous clue for adoptees.

Finding Biological Ancestors

An increasingly common use of Y-DNA is to recover an unknown biological surname. For an adopted male, for example, the Y-DNA retains a link to a biological family that paper records may not possess, or that might be locked behind privacy walls. Based on my experience with the program, roughly 30 percent of males who test their Y-DNA through the Adopted DNA Project <**www.familytreedna.com/public/adopted**> at Family Tree DNA are able to identify their likely biological surname through Y-DNA alone.

For example, assume an adoptee by the name of Riley Graham has done extensive research but has not found any accessible records that reveal his biological surname. In an attempt to connect with biological relatives, Riley takes a 67-marker test, and his results reveal an interesting pattern:

Genetic Distance	Name	Most Distant Ancestor	Y-DNA Haplogroup	Terminal SNP
0	Roger Davis	Joshua Davis, b. c. 1765 MD	R-L1	
0	Philip Davis	Joshua Davis, b. c. 1765 MD	R-L1	
1	Frederick Davis	Nathaniel Davis, b. 1772 MD	R-P25	P25
2	John Thomas		R-L1	

Riley shares all markers with Roger and Philip Davis, meaning he is closely related to these individuals on his patrilineal line. It is very likely, therefore, that his biological father, grandfather, or other recent ancestor had the Davis surname. Riley and Frederick Davis have a genetic distance of 1, so their relationship is potentially a little more distant. Roger and John Thomas have a genetic distance of 2. This could represent a non-paternal event. A non-paternal event can occur, for example, if a Thomas ancestor adopted a Davis child, if the wife of a Thomas ancestor had an affair with a Davis man, or if a Thomas male decided to change his surname to Davis.

Y-DNA testing will not always reveal the surname as it has in this example. Often, the individuals in the test-taker's match list will be too distantly related to be definitive. For example, the results may show a match list with several or many different surnames. Alternatively, few people may be in the match list, or they may all be very distantly related. In that instance, the test-taker can wait for other men (potential new matches) to take Y-DNA tests, or can identify men who might be good candidates and ask if they are willing to undergo Y-DNA testing.

CORE CONCEPTS: Y-CHROMOSOMAL (Y-DNA) TESTING

- The Y chromosome is one of the two sex chromosomes. Only men possess a Y chromosome, and a father passes down his Y chromosome to only his sons. As a result of this unique inheritance pattern, Y-DNA is only used to examine a test-taker's paternal line.

- Y-DNA testing is done by either sequencing short regions of the Y chromosome (Y-STR testing) or through SNP testing (Y-SNP testing) of the Y chromosome.

- The results of any Y-DNA test can be used to determine the paternal haplogroup, or ancient origins, of the paternal line back thousands of years. Y-STR tests estimate the paternal haplogroup, while Y-SNP tests definitively determine the paternal haplogroup.

- The results of an Y-STR sequencing test can be used to fish for genetic cousins. Since the Y chromosomes mutates relatively rapidly and at a well-characterized rate, Y-STR testing is very good at finding random genetic matches in a testing company's database and estimating how many generations have passed since two men shared a common paternal ancestor.

- Y-DNA test results can be useful for examining specific genealogical questions, such as whether two people are or are not paternally related.

DNA in Action

Are They Brothers?

In the diagram below, a genealogist has identified two historical men, Philip and Joseph, as potential brothers based on well-researched paper-trail evidence. To determine whether the two men could have been brothers, the genealogist has traced descendants of Philip and Joseph and asked them to take a Y-STR test. Both descendants agreed, and the genealogist is now reviewing the results.

The results of the 67-marker Y-STR test taken by the two descendants, a brief excerpt of which is shown below, reveal that the two test-takers are identical at all sixty-seven markers:

DYS#	393	390	19	391	385a	385b	426	388	439
Philip's Descendant	13	24	14	10	11	14	12	12	12
Joseph's Descendant	13	24	14	10	11	14	12	12	12

Although Philip's descendant and Joseph's descendant have identical Y-DNA test results, this does not alone prove that Philip and Joseph were brothers. Just like mtDNA, Y-DNA cannot determine an exact relationship, and thus the results only provide additional support for the hypothesis that Philip and Joseph *could* have been brothers. They also could have been father/son, uncle/nephew, paternal male cousins, or a variety of other possible relationships, as long as they share a paternal line. Indeed, they could even be very distant paternal cousins. Still, it is feasible to explore the possibility that Philip and Joseph could be brothers, particularly in view of documentary evidence.

However, let's assume that the results actually came back and indicated that they were not similar, or that they even belonged to completely different haplogroups. That would mean that at least one of the following scenarios is true: (1) Philip and Joseph were in fact not brothers or (2) somewhere in the patrilineal line between Philip and his purported Y-DNA descendant—or between Joseph and his purported Y-DNA descendant—a "break" in the line, such as an adoption, occurred.

As mentioned earlier in this chapter, a break in a Y-DNA line is known as misattributed parentage or a non-paternal event (NPE). NPEs occur at a rate of approximately 1 to 2 percent in the generation population, and can be due to a variety of factors such as adoption, name change, infidelity, and others. Although NPEs are rare, they should always be considered when reviewing Y-DNA test results.

DNA in Action

Is This the King? Part II

As discussed in the previous chapter, skeletal remains thought to be of King Richard III were found in 2012 under a parking lot in Leicester, England. The thirty-two-year-old king was killed in the Battle of Bosworth Field and buried within the Greyfriars Friary Church in Leicester, England. However, the location of Richard's grave was ultimately lost through the passage of time. DNA testing of the remains determined that the skeleton's relatively rare mtDNA was identical or nearly identical to the mtDNA of two very distant descendants of King Richard III's sister, Anne.

To further support the hypothesis that the skeletal remains were those of Richard III, the researchers wanted to compare the Y-DNA obtained from the skeleton to Y-DNA from some of Richard's paternal relatives. Since Richard III had no children, genealogists had to go back to Richard's great-great-grandfather, Edward III, and follow his descendants forward to find a candidate who would share Y-DNA with Richard III. The genealogists ultimately identified five living descendants who took Y-DNA tests (labeled A through E).

Y-SNP testing of individuals A through E revealed that four of them belonged to the Y-DNA haplogroup *R1b-U152* (a single patrilineal group). However, one of the individuals belonged to haplogroup *I-M170* and thus was not a patrilinear relative of the other four within the time span considered, indicating that a break had occurred within the last four generations. In contrast, the Y-DNA sequenced from the skeletal remains belonged to haplogroup *G-P287*, with a corresponding Y-STR haplotype. Thus, surprisingly, the Y-DNA of the two brother's lines (John of Gaunt's line and Edward of York's line) does not match.

Overall, the evidence (such as the Y-DNA results and the results of the mtDNA test, mentioned in chapter 4) overwhelmingly concluded that the skeletal remains were indeed those of Richard III. The Y-DNA results also suggest there is a case of misattributed parentage somewhere between Richard III and individuals A through E. Because four of the test-takers had the same Y-DNA, the case of misattributed parentage is almost certainly at or before Henry Somerset. Nineteen generations separate Richard III and Henry Somerset, and, assuming a rate of NPEs of 1 to 2 percent per generation, the chance of misattributed parentage occurring in this number of generations is 16 percent.

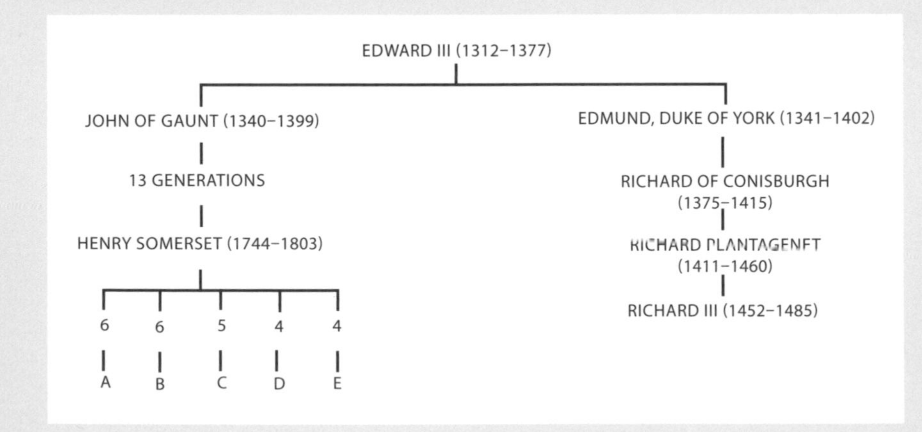

6

Autosomal-DNA (atDNA) Testing

You sent away a saliva sample or cheek swab to one or more of the major testing companies (23andMe <**www.23andme.com**>, AncestryDNA <**www.dna.ancestry.com**>, and Family Tree DNA <**www.familytreedna.com**>) for an autosomal-DNA (atDNA) test, and you just received your results. What do you do now? What do these results mean, and how do you use them to advance your genealogical research?

Within just the past few years, several million people have taken atDNA tests at 23andMe, AncestryDNA, and Family Tree DNA. And with more people taking these tests and entering into companies' databases, it's easier than ever before to find genetic matches and search for common ancestors. In this chapter, we'll review the fundamental concepts needed to understand atDNA test results and some of their functions, such as cousin matching. We'll also review the atDNA tools offered by the testing companies, and how to use these tools to find common ancestors and answer genealogical questions.

What is Autosomal DNA?

Autosomal DNA refers to the twenty-two pairs of non-sex chromosomes found within the nucleus of every cell. The atDNA chromosomes, or **autosomes**, vary in length, and

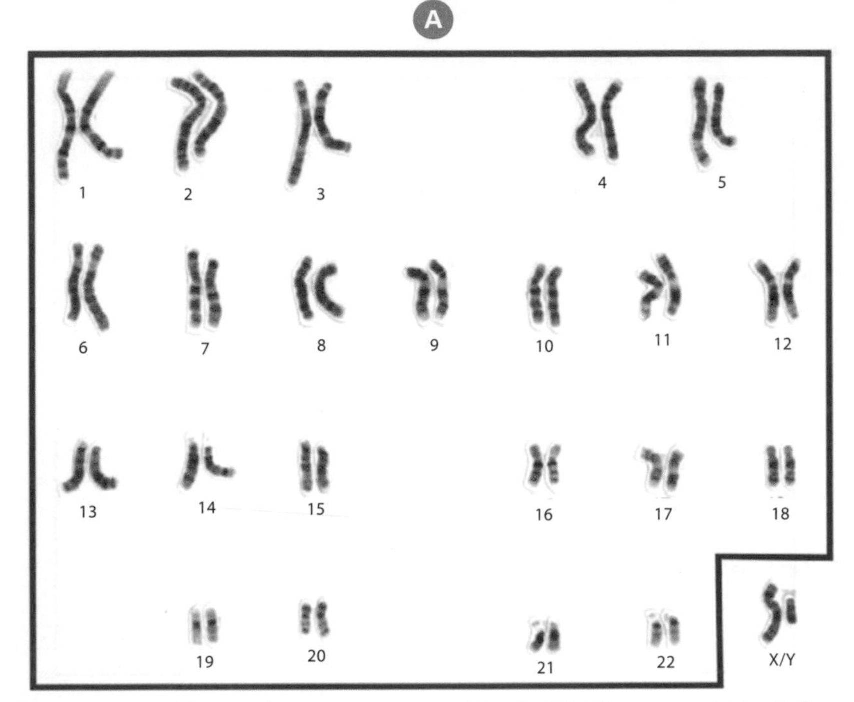

Twenty-two sets of human chromosomes are considered atDNA. The twenty-third pair, the sex chromosomes, determine gender, among other traits.

when they are visualized (such as in image A), they are numbered approximately in relation to their sizes, with autosome 1 being the largest and autosome 22 being the smallest.

atDNA Inheritance

atDNA, unlike mitochondrial DNA (mtDNA) and Y-chromosomal DNA (Y-DNA), is inherited equally from both parents. Accordingly, an individual gets one chromosome in each chromosome pair from Mom, and one chromosome in each chromosome pair from Dad (image B). Unfortunately, since the chromosomes are not labeled or marked in a way that easily identifies which parent they come from, the results of a single atDNA genetic genealogy test cannot identify the specific source of a piece of DNA.

A child inherits his entire DNA from his parents, about 50 percent of his DNA from his mother and about 50 percent of his DNA from his father. However, the child is not inheriting all of his parents' DNA; instead, he's only inheriting half of his parents' *total* DNA and leaving half of it behind. This occurs at every generation, meaning that as we go back in time, we inherit less DNA from ancestors at each generation.

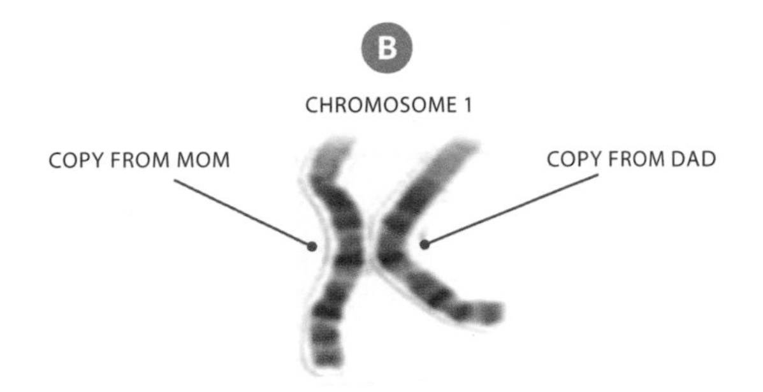

You receive one copy of each chromosome from each parent.

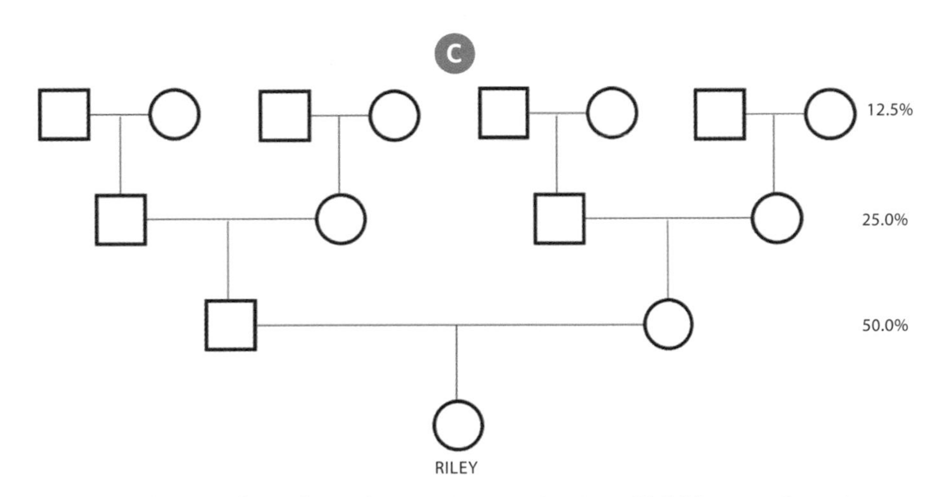

Less ancestral DNA is inherited at each generation, meaning that atDNA "disappears" over time.

In image C, Riley inherits only half of his parents' DNA, 25 percent of his grandparents' DNA, and just 12.5 percent of his great-grandparents' DNA. Although not shown in the image, Riley will inherit just 6.25 percent of his great-great-grandparents' DNA, and so on.

It should be noted that the percentages are averages across a population instead of absolutes for any given individual. Thus, while *on average* an individual will inherit 25 percent of his DNA from each grandparent, in practice the percentages will vary. For example, here is a graph of observed percentages of DNA received from four grandparents for two sibling grandchildren:

	Paternal Grandfather	Paternal Grandmother	Maternal Grandfather	Maternal Grandmother
Expected	25.0%	25.0%	25.0%	25.0%
Grandson 1	28.0%	22.0%	26.6%	23.4%
Grandson 2	23.7%	26.3%	17.7%	32.3%

Although each will average 25 percent, the range for Grandson 1 is 22.0 to 28.0, and the range for Grandson 2 is larger: 17.7 to 32.3.

Due to the inheritance pattern seen above, it is possible to determine how much DNA an individual is likely to share with close relatives. For example, if a grandchild and a grandparent both take an atDNA test, they should share *on average* about 25 percent of their DNA. Similarly, if an individual and his aunt both take an atDNA test, they should share on average about 25 percent of their DNA.

Image D shows how much DNA, in percentages, an individual is predicted to share with close relatives. The percentage for each relationship can be found in the red boxes. As with the other percentages above, this chart only represents the average percentage

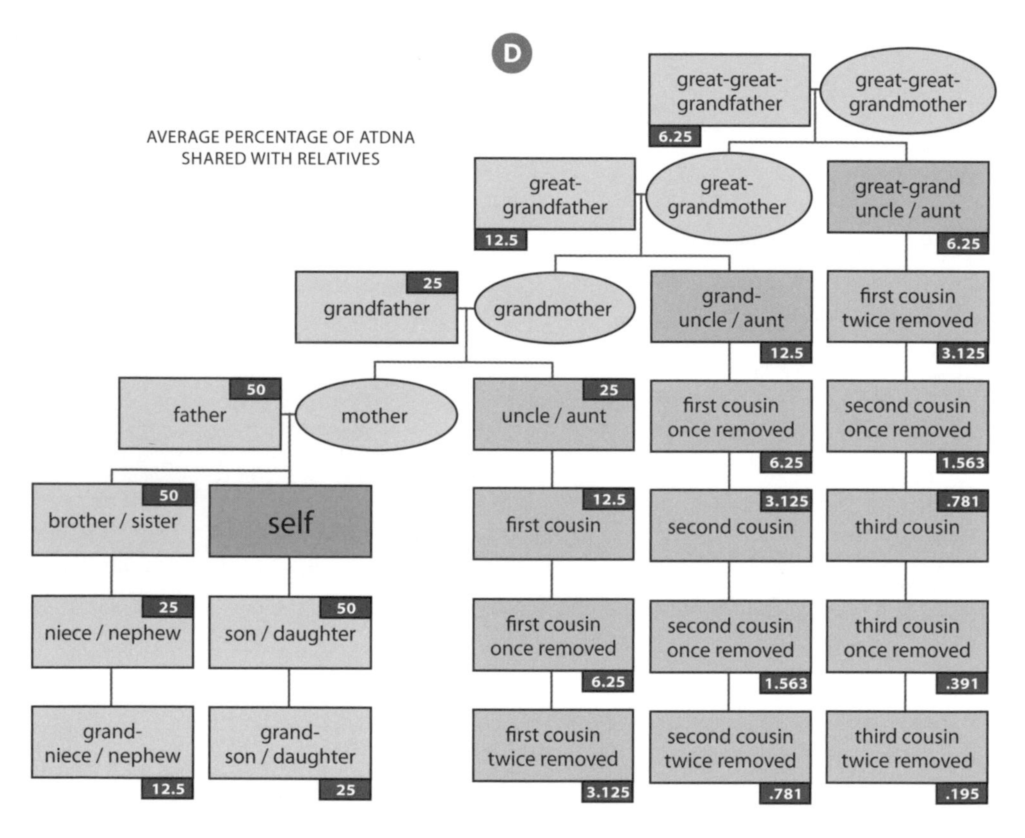

You can predict what percentage of DNA you're likely to share with relatives based on atDNA inheritance patterns.

of DNA shared with relatives. The actual amount of DNA shared with a relative can vary quite a bit.

Recombination

One important factor to consider when taking atDNA tests and interpreting results is the process of **recombination**. Before a chromosome is passed down to the next generation, it undergoes recombination, in which a parental chromosome pair optionally exchanges pieces of DNA during **meiosis**, a natural, specialized process in which cells divide as eggs and sperm are created for reproduction.

Before we dig into recombination, it might be helpful to review meiosis as a whole and how and when recombination can occur. Meiosis occurs so that cells can divide their DNA amongst their daughter cells during gamete (i.e., sperm or egg) production, and the cell duplicates its chromosomes very early in meiosis. Normally, every cell has twenty-three pairs of chromosomes (twenty-two pairs of autosomes and one pair of sex chromosomes), for a total of forty-six chromosomes. However, in the first step of meiosis, the chromosomes are duplicated, resulting in a total of ninety-two chromosomes. Using image E as an example, the cell will duplicate its DNA so it has four copies of

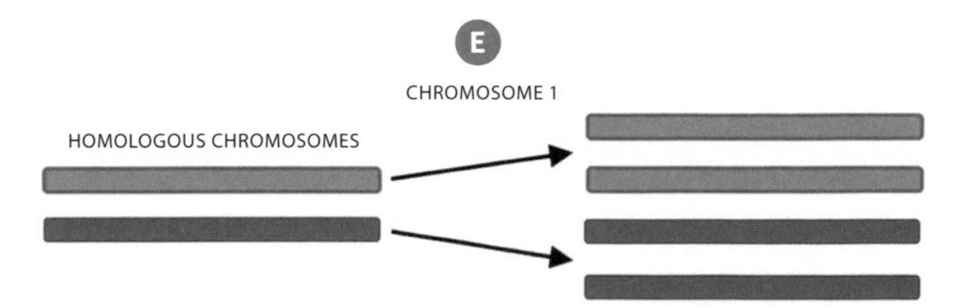

During meiosis, each pair of chromosomes (one copy from the mother and one from the father) duplicates itself, producing four total copies of each chromosome. In this image, DNA inherited from the father is in blue and DNA from the mother in pink.

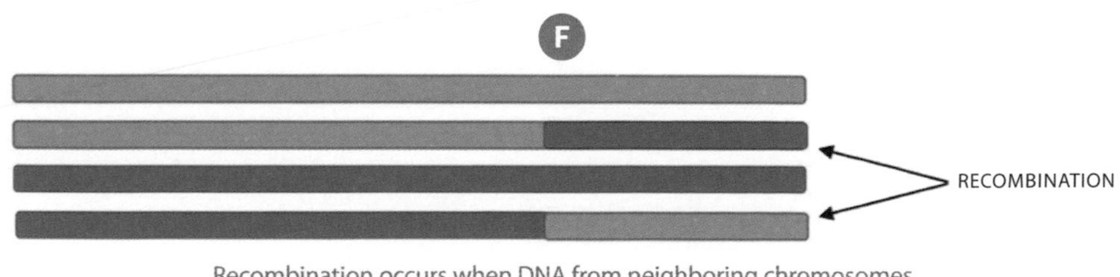

Recombination occurs when DNA from neighboring chromosomes cross over and exchange genetic information.

chromosome 1 (two copies of the chromosome from the person's mother, and two copies of the chromosome from the person's father). Similarly, the cell will have four copies of chromosome 2, and so on.

As the now-duplicated chromosomes line up to be split into daughter cells, recombination can occur between any of the four copies of a chromosome (such as chromosome 1) as strands of chromosomes overlap. Should chromosomes' genetic material cross over, some DNA may be exchanged between them, possibly resulting in a genetic variance. And once meiosis (and any recombination event/events) is complete, the daughter cells will randomly receive *just one* of the four chromosome copies, meaning three copies (two of them identical) will be left behind.

Note that recombination events may or may not be detectable, based (in part) on what chromosomes crossed over genetic information. If recombination occurs between the two paternal copies of the chromosome or between the two maternal copies of chromosome 1 (between **sister chromatids**—that is, between the two blue paternal chromosomes or between the two pink maternal chromosomes), there is no detectable change because they are identical copies. However, if recombination occurs between a paternal and a maternal chromosome (between **nonsister chromatids**—that is, between a blue paternal chromosome and a pink maternal chromosome), a detectable crossover occurs (image F).

Recombination happens randomly, and each cell division can result in a different amount of recombination events (or no recombination events at all). It's unusual for more than a handful of recombination events for a single chromosome to occur, however. Interestingly, females tend to have more recombination events over the entire set of twenty-two autosomes than males do.

The following example (image G) demonstrates the passage of atDNA from a paternal grandmother (Agatha) to her granddaughter (Courtney). This represents a single recombination event, when the father (Benny) created the sperm. (Note: Although

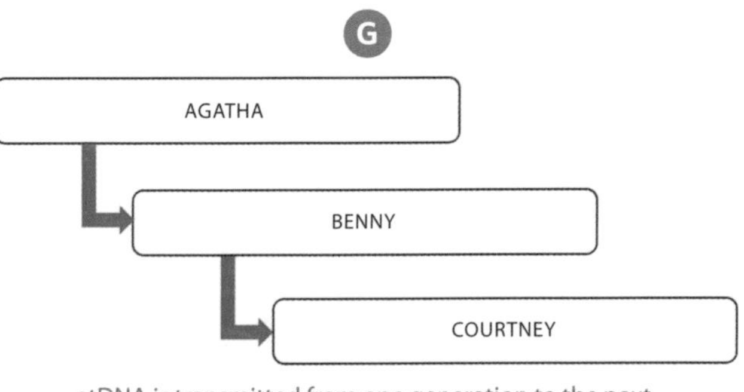

atDNA is transmitted from one generation to the next, regardless of each family member's gender.

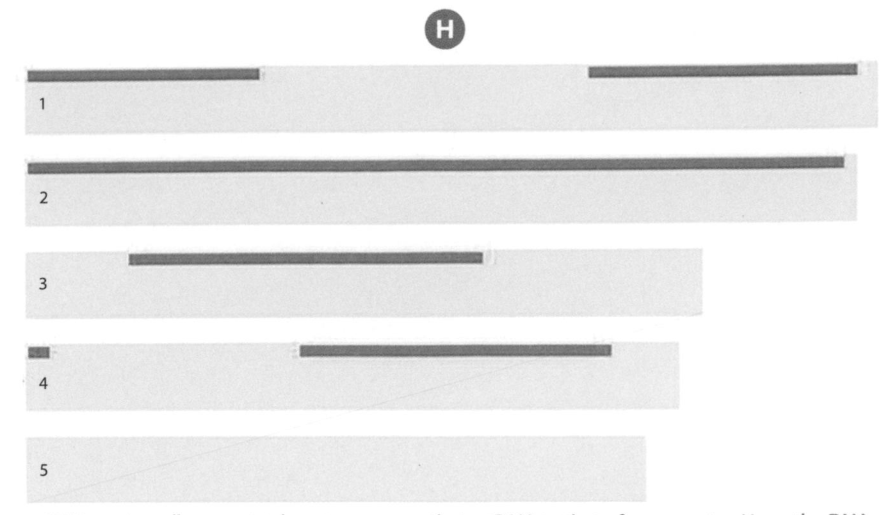

atDNA testing allows test-takers to compare their atDNA to that of an ancestor. Here, the DNA that Courtney shares with her grandmother, Agatha, on her first five chromosomes is in green.

Agatha's DNA underwent recombination when she created the egg that would become her son Benny, that recombination would only be detected by comparing her DNA to her ancestors.) Image H compares Courtney's first five chromosomes to Agatha's first five chromosomes, with the DNA shared by both in green.

Comparing the two women's atDNA can tell us a lot about how atDNA is inherited and recombined across generations, as recombination must have occurred in regions where the women don't share DNA. Chromosome 1, for example, suggests there were two recombination events at each place on the chromosome where the two women differ, indicated in image I. There may also have been a third recombination at the end of chromosome 1, as the women don't share that region as well.

So how do we explain the differences between them, and where did the recombination events occur? Before the father passed down a copy of chromosome 1 to his daughter, his maternal and paternal copies of chromosome 1 crossed over at least two different locations. And when Benny's DNA was split up into his daughter cells, the copy with genetic material matching his mother's at the locations in green was passed on to the cell that became Courtney. (Note: The other copy of the chromosome would look just the opposite when compared to Agatha's DNA—with the shared green segment in the middle. However, Courtney didn't inherit this copy of the chromosome.)

The shared DNA also helps researchers determine which part of the chromosome was inherited from other ancestors. Specifically, since the segment in the middle of chromosome 1 doesn't match Agatha's, it *must* match Courtney's paternal grandfather (i.e., Benny's father); Benny's DNA, like Courtney's, can only come from two of his

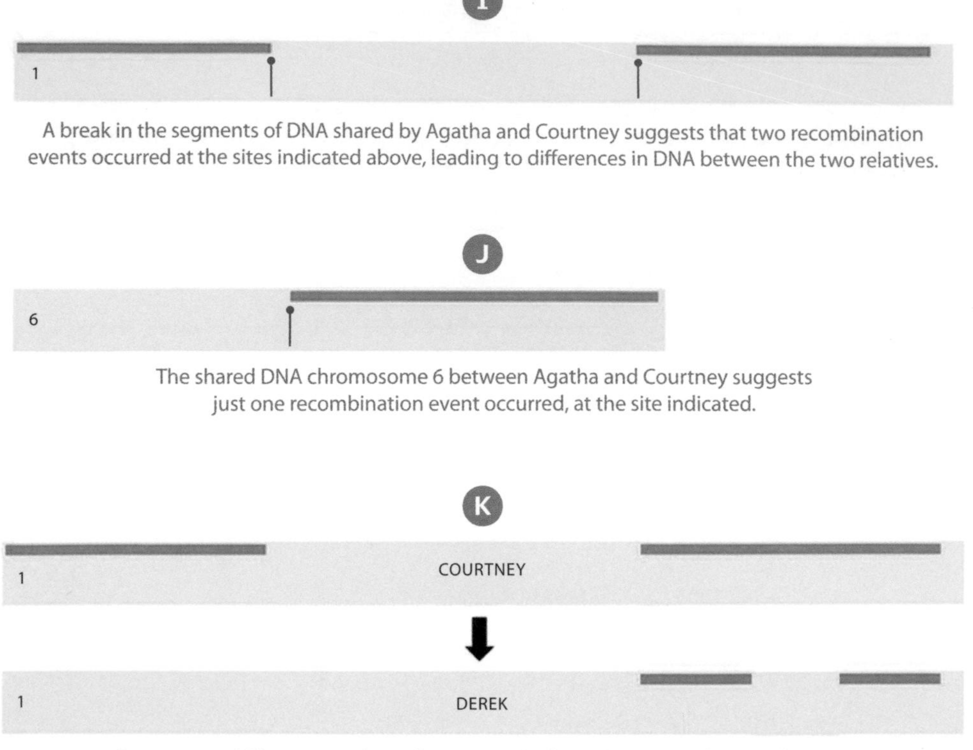

A break in the segments of DNA shared by Agatha and Courtney suggests that two recombination events occurred at the sites indicated above, leading to differences in DNA between the two relatives.

The shared DNA chromosome 6 between Agatha and Courtney suggests just one recombination event occurred, at the site indicated.

Courtney could have passed Agatha's atDNA (indicated in green) down to her son, Derek, in a number of ways, including the above.

ancestors: his mother, Agatha (Courtney's paternal grandmother), or his father (Courtney's paternal grandfather).

Let's look at some more chromosomes and see what we can assume from them. On chromosomes 2 and 5, there was no recombination between nonsister chromosomes, and Courtney inherited Agatha's *entire copy* of chromosome 2, but none of Agatha's chromosome 5. By process of elimination, Courtney must have instead inherited her paternal grandfather's entire copy of chromosome 5. On chromosome 6, we know there was a single recombination event about halfway down the chromosome, as Agatha and Courtney don't share a significant portion of chromosome 6's DNA (image J).

Importantly, Agatha's DNA that wasn't passed on due to recombination is now lost to Courtney and all future generations (unless it comes back in from other lines/through Benny's siblings). For example, none of Courtney's descendants will inherit DNA from Agatha's chromosome 5. Thus, all the genes, ethnicity markers, and other information contained on the Agatha's copy of chromosome 5 are lost in this particular

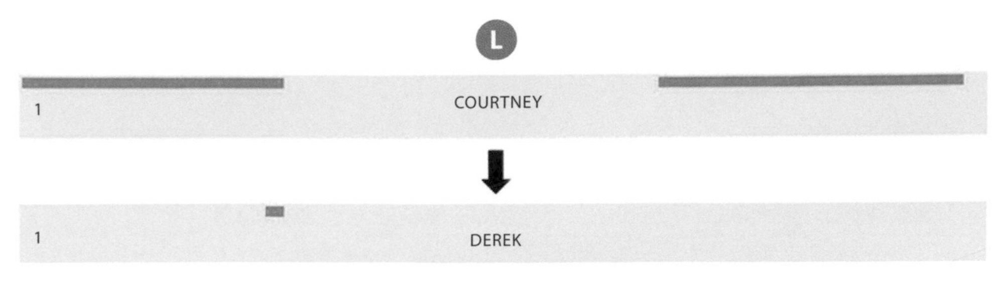

Courtney may have only passed down a small amount of Agatha's atDNA to Derek.

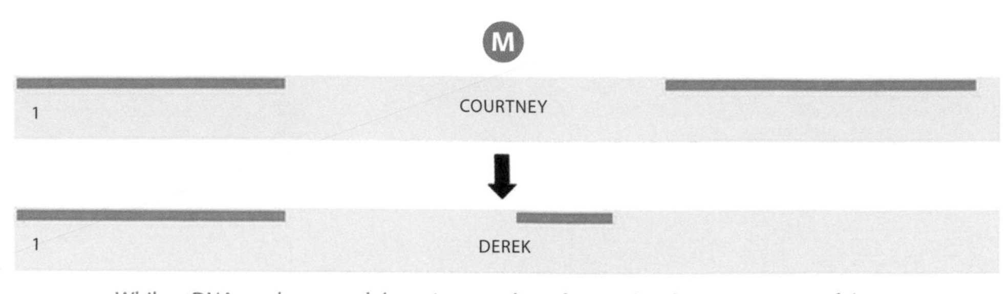

While atDNA can be passed down in a number of ways, the above is not one of them. Courtney cannot pass down atDNA from Agatha that she herself doesn't have. The only way for this to occur is if Derek's father is also somehow related to Agatha.

line of the family (although, again, it may be recovered by testing other relatives/other of Agatha's descendants).

Note that, because the recombination of DNA can occur at every generation, the amount of DNA between an ancestor and her descendants often shrinks from one generation to the next. For example, Agatha's chromosome 1 DNA that was passed on to Courtney will potentially be further "broken down" into smaller pieces when Courtney passes it on to her children. Courtney's chromosome 1 DNA could recombine when producing the cells that become her son, Derek. In image K, in which the chromosome 1 DNA shared with Agatha by Courtney (top) and Derek (bottom) is indicated in green, the large segment of Agatha's DNA at the left end of chromosome 1 was not passed on to her great-grandson, and the large segment at the right end of chromosome 1 experienced two recombination events that resulted in further loss. Alternatively, recombination could result in only a very small piece of Agatha's chromosome 1 being passed to the next generation, as it does in image L.

As stated earlier in this section, DNA that isn't passed down from one generation to the next can't be passed down to future generations. As a result, Derek and Agatha could never share the portions of chromosome 1 pictured in image M, as Courtney never inherited that segment of DNA from her father, and so could never pass it down to Derek. The only exception to this rule would be if Derek somehow inherited that segment of Agatha's

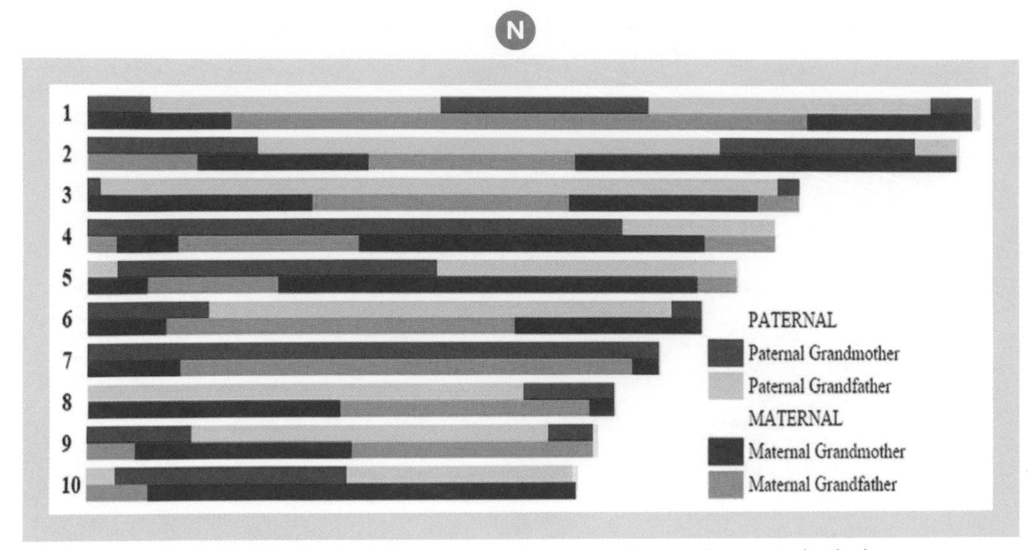

Tools like Kitty Cooper's Chromosome Mapper can illustrate how an individual inherits atDNA from ancestors, such as the above, which indicates which sections of atDNA a grandchild received from each of his grandparents.

chromosome 1 DNA from his father, which would likely mean that Derek's father is also somehow related to Agatha since he shares some of her DNA.

Now, let's pan away to a real-life example: a grandchild's DNA is compared to all four of his grandparents, with the source of every piece of DNA (for the first ten chromosomes) identified (image N). The graph, which was made using Kitty Cooper's Chromosome Mapper <**www.kittymunson.com/dna/ChromosomeMapper.php**>, shows how each section of the grandchild's DNA compares to each of his grandparents: Maternal grandparents are in red and orange, and paternal grandparents are in dark and light blue. Recombination events took place wherever colors change. Along each paternal chromosome (dark and light blue), for example, the grandchild matches either the paternal grandmother (dark blue) or the paternal grandfather (light blue). According to this graph, there were between zero and four recombination events per paternal or maternal chromosome. On chromosome 7, for example, the grandchild received an entire copy of the paternal grandmother's chromosome, meaning that he shares no DNA with his paternal grandfather on that particular chromosome.

Two Family Trees

As stated in chapter 1, genealogists must actually consider two, distinct family trees when conducting genetic research. The first, the genealogical family tree, contains every parent, grandparent, and great-grandparent back through history. This is the tree that genealogists spend their time researching, often using paper records such as birth and death certificates, census records, and newspapers to fill in ancestors and information about them. The second family tree is the genetic family tree, a subset of the genealogical family tree that contains only those ancestors who contributed to the test-taker's DNA. Not every person in the genealogical family tree contributed a segment of their DNA sequence to the test-taker's genome. In fact, the genetic tree is only guaranteed to contain both biological parents, each of the four biological grandparents, and each of the eight biological great-grandparents, but with each generation it is much less likely that every person in that generation contributed a piece of their DNA to the test-taker's DNA.

The difference between the two trees results in facts important to consider when tracing atDNA inheritance, including:

- **Siblings have different genetic family trees.** Other than identical twins, full siblings share only about 50 percent of their DNA (and half-siblings share about 25 percent of their DNA in common). As a result, the siblings have many genetic ancestors in common, but there are many distant ancestors represented in one sibling's DNA that are not represented in the other sibling's DNA. While full siblings have the same genealogical family tree, they have differing genetic family trees.
- **Genealogical cousins are not always genetically related.** First cousins share a strong genealogical and genetic link. They both are descended from shared grandparents, and both inherit some of the same DNA from those shared grandparents. However, it is much less likely that fifth cousins share DNA in common, since one or both of them may not have inherited DNA from the shared ancestor. Indeed, there is roughly only a 10- to 30-percent probability that fifth cousins will share a common DNA segment from their shared ancestor(s).
- **Ethnicity is nearly impossible to predict.** One of the most popular uses of atDNA testing is to estimate an individual's ethnic heritage (also called "ethnicity" or "biogeographical estimate"). Chapter 9 is devoted to this use. However, since an individual does not possess *all* the DNA of his ancestors, he does not necessarily represent the entire ethnicity of his ancestors.

How the Test Works

The atDNA tests currently offered by 23andMe, AncestryDNA, and Family Tree DNA are SNP tests, meaning that they sample hundreds of thousands of SNPs—the variable nucleotides *A*, *T*, *C*, and *G*—located throughout the twenty-two autosomal chromosomes. Although sequencing all of an individual's DNA (called **whole-genome sequencing**) will soon be as affordable as SNP testing, the higher price has prevented it from being offered commercially by these companies. In the future, genetic genealogists are likely to purchase whole-genome sequencing instead of SNP testing.

When the testing company receives the saliva sample from the test-taker, it extracts the DNA and makes many copies. The company then uses the test-taker's amplified DNA to test for the nucleotide value at each of seven hundred thousand or more locations within the test-taker's DNA. The results of the test will most often look like this:

rsID	Chromosome	Position	Result
rs3094315	1	752566	AA
rs12124819	1	776546	AG

Each line of the table represents an SNP somewhere in the test-taker's genome. "rsID" stands for Reference SNP cluster ID, and is a general reference for an SNP. The "chromosome" and "position" columns reveal where within the genome the result is found. The "result" column is the value of the maternal and paternal chromosomes at that location. Without more information, however, it's not possible to determine which result is the paternal chromosome and which result is the maternal chromosome.

As we'll see throughout this chapter, the results of an atDNA test can have several important uses. For example, the results are most often used for finding genetic relatives, people who share a segment of DNA with the test-taker.

Using atDNA: Finding Genetic Cousins

In addition to using DNA to break down genealogical brick walls, researchers often use DNA to find cousin matches amongst other test-takers. And while many testing companies do the heavy lifting for you, you'll still have to consider a number of factors when attempting to find and confirm genetic cousins. In this section, we'll discuss some of the factors that influence whether or not two people are genetic cousins and help you analyze DNA cousin results each testing company provides.

Minimum Segment Length

Each of the testing companies has selected a minimum segment length threshold that must be met before two people in the testing database will be flagged as sharing DNA in common, and this threshold can be key to understanding your results. If the threshold is set too low, some of the individuals identified by the company will be false positives, meaning that the two test-takers are either not actually related or have a shared common ancestor who lived thousands of years ago. If the threshold is set too high, there can be false negatives, meaning that the test-takers share enough DNA segments to be true genetic relatives but are being arbitrarily excluded from the test-taker's list of people sharing their DNA.

Ideally, the testing company would like to identify only those individuals who share a common ancestor within the past three to four hundred years (what may be called "a genealogically relevant time frame") while excluding individuals who share a common ancestor more than five hundred years ago. While the minimum segment threshold helps this goal, it is not perfect.

23ANDME

At 23andMe, two individuals are identified as a genetic match if they share at least one segment of at least 7 centimorgans (cMs) *and* seven hundred single nucleotide polymorphisms (SNPs). Additional segments beyond the initial 7-cM segment are identified as being shared by the two individuals if those segments share at least 5 cMs and seven hundred SNPs. Thus, if the results of two people at 23andMe only show that they share a single segment of 6.5 cMs and 750 SNPs, they will not be identified as a genetic match, since there is no segment of at least 7 cMs shared by the two individuals.

For the X chromosome at 23andMe, there are different thresholds depending on the sex of the two test-takers:

- Male versus male: 1 cM and two hundred SNPs
- Female versus male: 6 cM and six hundred SNPs
- Female versus female: 6 cM and twelve hundred SNPs

Notably, 23andMe has a set cap of approximately two thousand genetic cousins for each test-taker. The two-thousand cap means that for many test-takers, valid matches are excluded from the list of genetic cousins. According to one estimate by the International Society of Genetic Genealogy (ISOGG) <www.isogg.org/wiki/Identical_by_descent>, the two-thousand cap excludes new matches below approximately 17 cM shared for most individuals with Colonial American ancestry.

ANCESTRYDNA

At AncestryDNA, two individuals are identified as a genetic match if they share at least one segment of at least 5 cMs. This is a relatively low threshold, and this threshold increases the probability that individuals identified as distant matches at AncestryDNA are actually false positives (which, again, would indicate they are related well beyond a genealogically relevant time frame).

FAMILY TREE DNA

At Family Tree DNA, two individuals are identified as a genetic match if they share at least one segment of at least 5.5 cMs, according to the company's website

<**www.familytreedna.com/learn/autosomal-ancestry/universal-dna-matching/genetic-sharing-considered-match**>. Other evidence suggests the threshold might be approximately 7.7 cMs and at least five hundred SNPs for the first segment, and a total of at least 20 cMs shared in common (including all the shorter matching segments between 1 cM and 7 cMs).

For the X-chromosomal (X-DNA) test at Family Tree DNA, the criteria is twofold: The individuals must already meet the atDNA threshold, and they must share a segment of at least 1 cM and five hundred SNPs. The requirement for a shared atDNA segment prior to X-DNA comparison means there are false negatives; individuals who share X-DNA but don't share atDNA will not be identified as genetic cousins. Further, the very low threshold of 1 cM and five hundred SNPs for X-DNA matching means that test-takers will have

SHARING DNA WITH SIBLINGS

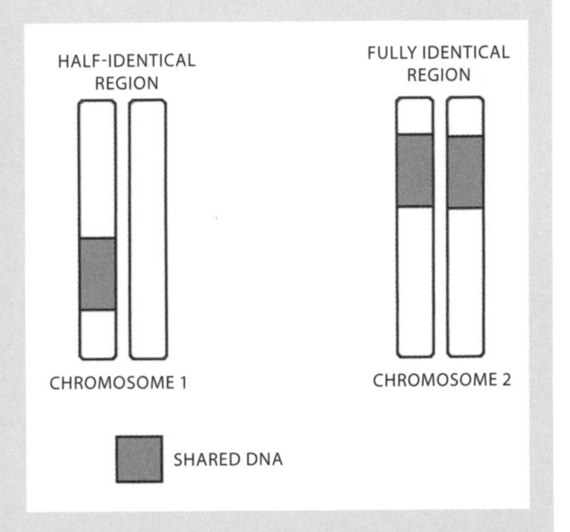

You might think that DNA comparisons between siblings would be straightforward, but (like with many topics in genetic genealogy) the answer is more complicated.

How much DNA is shared by siblings raises an important distinction in genetic genealogy. Individuals who share DNA can be either half-identical or fully identical. A **half-identical region** (HIR) is a portion of the genome where two people share a segment of DNA on just one of their two chromosomes. Remember, everyone has two copies of each chromosome, and it is possible to share DNA with someone else on one of those copies, or in rarer cases (such as siblings) on both of those copies. Accordingly, a **fully identical region** (FIR) is a region where two people share segments of DNA on *both* copies of their two chromosomes. The image demonstrates HIR versus FIR shared segments for an individual, where the blue segment(s) is the DNA shared by that test-taker with another test-taker

Half of the DNA shared by siblings is from HIRs (1700 cM), while half of the DNA shared by siblings is from FIRs (850 cM, for a total of 1,700 cM). The HIRs plus the FIRs equals 3,400 cMs. However, 23andMe and Family Tree DNA only report half of the DNA at FIRs. Accordingly, these companies will only report about 75 percent of the actual amount of DNA shared by siblings (or, in other words, just 75 percent of 50 percent, or 75 percent of 3,400 cMs).

In summary, siblings are half-identical on 50 percent of their DNA (1,700cM) and fully identical on a further 25 percent/850cM.

false positives—individuals who are identified as sharing X-DNA within a genealogically relevant time frame, but likely do not.

Although these matching thresholds are in place to maximize the likelihood that individuals identified as sharing a segment of DNA are in fact recent genetic cousins, it is important to keep in mind that every "match list" will have individuals who are false positives. Accordingly, it is often a best strategy to focus on those individuals who share the most DNA. The longer a shared segment is, and the more segments two people share in common, the greater the likelihood that the two individuals share a recent common ancestor.

Likelihood of Sharing DNA

As discussed earlier, only a small percentage of genealogical cousins actually share DNA. After seven to nine generations, DNA is not inherited by all descendants of an ancestral couple. Further, the *same* pieces of DNA are not inherited by all descendants of an ancestral couple, even at the first generation. In other words, a great-great-great-grandson may have inherited the only piece of DNA from his great-great-great-grandfather on chromosome 8, while the only piece of DNA a great-great-great-granddaughter inherited from that same great-great-great-grandfather is on chromosome 3. Although these two individuals are fourth cousins and both have DNA from their shared ancestors, they do not share any segments of DNA in common. To use terminology from chapter 1 and an earlier section of this chapter: They are genealogical cousins, but not genetic cousins.

What is the likelihood that genealogical cousins will share DNA? For close cousins, the likelihood is very high, but it decreases rapidly. All three companies have provided their estimate or calculation of these probabilities:

	23andMe	**AncestryDNA**	**Family Tree DNA**
	<customercare.23andme.com/hc/en-us/articles/202907230-The-probability-of-detecting-different-types-of-cousins>	<dna.ancestry.com/learn>	<www.familytreedna.com/learn/autosomal-ancestry/universal-dna-matching/probability-relative-share-enough-dna-family-finder-detect>
Closer than a Second Cousin	~100%	100%	> 99%
Second Cousin	>99%	100%	> 99%
Third Cousin	~90%	98%	> 90%
Fourth Cousin	~45%	71%	> 50%
Fifth Cousin	~15%	32%	> 10%
Sixth Cousin	<5%	11%	< 5%

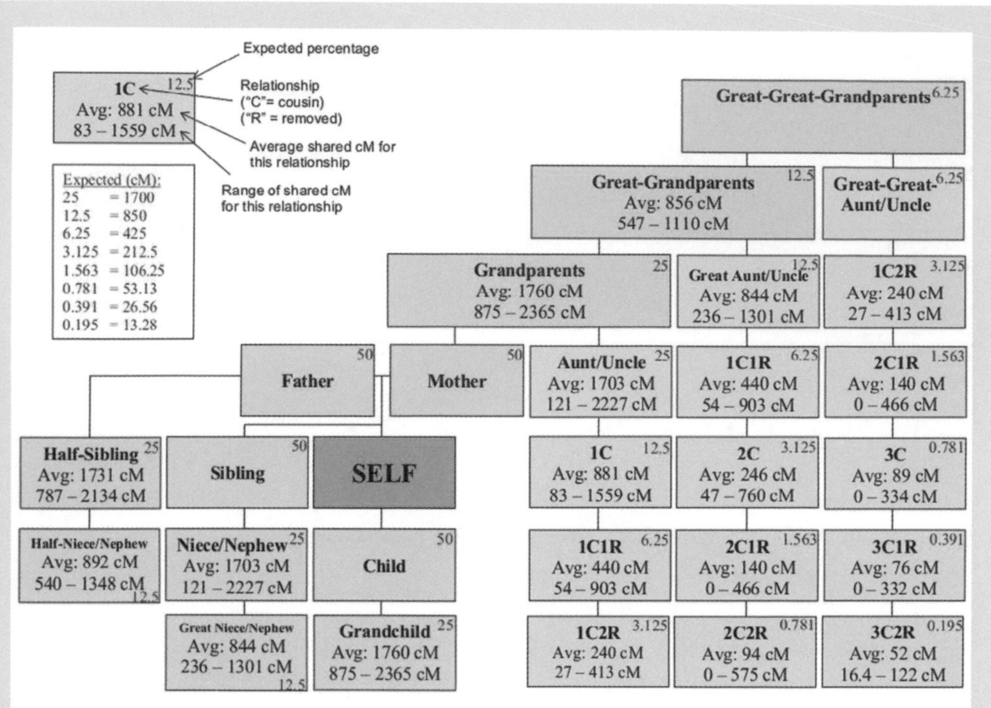

Blaine T. Bettinger – CC 4.0 Attribution License (2015)

THE SHARED CM PROJECT

The Shared cM Project **<www.thegeneticgenealogist.com/2015/05/29/the-shared-cm-project>** is a data collaboration project I started in 2015 to collect shared DNA data for known genealogical relationships to the third-cousin level. Although tables available at the time (and reproduced in this chapter) show how much shared DNA can be *expected* for these relationships, there was no good source of information about how much shared DNA is *actually observed* for these relationships. Accordingly, the project requested that genealogists submit data about their genealogical relationship, including how much DNA they shared with a cousin in total, and the largest segment they shared with the cousin. More than six thousand relationships were submitted, and the information was collated into tables and the following image. For each relationship, the following information is provided based on the data submitted for the relationship: (i) the average amount of DNA shared; (ii) the lowest amount shared; and (iii) the highest amount shared.

For example, those with a 1C1R (first cousins once removed) relationship are predicted to share 6.25 percent, or 425 cM, of their DNA. According to the data submitted to the Shared cM Project (which according to the table consisted of 606 different 1C1R relationships), first cousins once removed share an average of 440 cMs, with the lowest reported amount being 54 cMs and the highest reported amount being 903 cMs.

According to these estimates, it is almost guaranteed that relatives at a second-cousin level and closer will share detectable amounts of DNA. Indeed, I've never heard of a confirmed case in which second cousins did *not* share DNA. This means that if second cousins take an atDNA test and they don't share DNA, there was almost certainly a misattributed parentage event. For more information, see my article "Are There Any Absolutes in Genetic Genealogy?" **<www.thegeneticgenealogist.com/2015/04/13/are-there-any-absolutes-in-genetic-genealogy>**.

Amount of Shared DNA

The amount of DNA shared by two people can also help determine the genealogical relationship between those two people, although it is not a perfect predictor. For example, if two test-takers share 1,500 cM of DNA in common, their relationship is likely a grandparent/grandchild, aunt/uncle or niece/nephew, or half-sibling relationship. However, if two test-takers share 75 cM of DNA in common, it will not be clear whether the match is a third cousin, second cousin once removed, or a more complicated relationship (e.g., double cousin). Relationship prediction typically works best when the relationship is a third cousin or closer.

The following chart, adapted from the ISOGG wiki page "Autosomal DNA Statistics," **<www.isogg.org/wiki/Autosomal_DNA_statistics>**, provides the *expected* amount of DNA shared between people having the identified relationship:

Percentage	cMs Shared	Relationship
50%	3400.00	Parent/child
50%	2550.00	Siblings (see the Sharing DNA with Siblings sidebar)
25%	1700.00	Grandfather, grandmother, aunt/uncle/niece/nephew, half-siblings
12.5%	850.00	Great-grandparent, first cousin, great-uncle/aunt, half-uncle/aunt
6.25%	425.00	First cousin once removed
3.125%	212.50	Second cousin
1.563%	106.25	Second cousin once removed
0.781%	53.13	Third cousin

It's important to remember that without other information, it is impossible to tell whether two people share DNA because of a genealogical relationship in one line, two lines, or multiple lines. In the following example, two people share three segments of DNA totaling 58.2 cMs, and the testing company predicts them to be third cousins.

Segment	Chromosome	Start	End	cMs
1	3	10725423	18905001	9.5
2	11	7561324	25779385	30.1
3	14	5037045	6709246	18.6

However, one test-taker has tested both parents and sees that Segment 2 in the table is shared with the test-taker's mother, while Segments 1 and 3 are shared with the test-taker's father. Thus, the test-taker is likely related to the other individual a bit more distantly, but through multiple different lines. However, if the test-taker hadn't also tested his parents, it would be much more challenging to determine the exact nature of this relationship, and it would have been easy to assume that the individual was actually a more recent, third cousin.

Chromosome Browsers

Family Tree DNA and 23andMe are the only testing companies that offer a **chromosome browser,** a tool that allows test-takers to see exactly what segments of DNA they are identified as sharing with another person. Chromosome browsers can provide more detail than can information about shared cMs and SNPs. However, each company's chromosome browser looks different and can be used in slightly different ways.

FAMILY TREE DNA

At Family Tree DNA, a test-taker can use the chromosome browser tool to look at shared segments with any individual with whom she is predicted to share DNA (and thus are shown in the "Family Finder—Matches" list).

Image **O** shows the first five chromosomes in the chromosome browser at Family Tree DNA, with the DNA shared by a set of first cousins. The dark blue shapes represent each chromosome, and the full image includes all chromosomes from 1 to 22. Each of the orange blocks represents a shared segment of DNA. Note that the segments are *not* perfectly to scale in the chromosome browser, so evaluating the size of segments based on visual appearance alone can be misleading. In addition to showing shared segments on chromosome 1 to 22, Family Tree DNA also shows shared segments on the X chromosome.

When the test-taker hovers the mouse or pointer over a segment, he'll see a pop-up box that shows the chromosome number, the start location (e.g., position 53,624,479), the stop location (e.g., position 96,298,324 in the image), and the total size of that segment.

All of the information about shared DNA, including the chromosome number and start and stop locations for each segment, can be downloaded to a spreadsheet. And downloading the info into a spreadsheet will reveal the same information for all of the

O

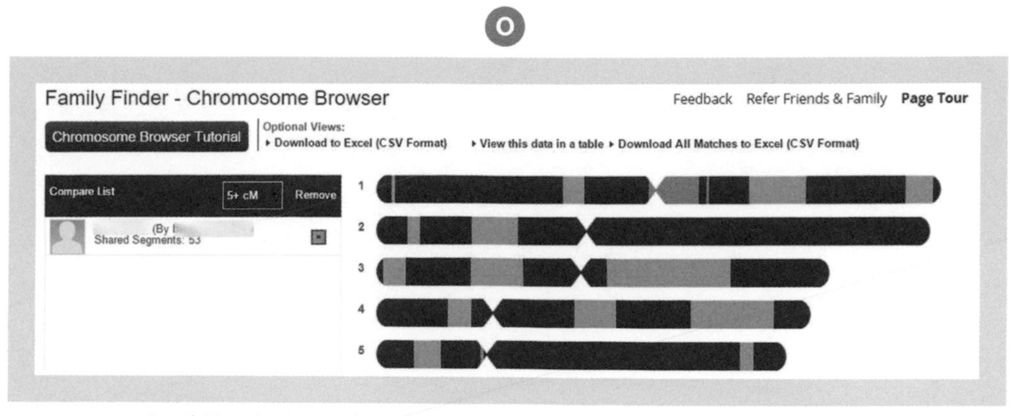

Family Tree DNA provides users with a chromosome browser to allow you to see your DNA in more detail, breaking down your shared DNA by chromosome.

shared segments. The following example provides a selection of just some of the segments shown in the first Family Tree DNA cousin comparison:

Chromosome	Start Location	End Location	cMs	No. of Matching SNPs
1	165402360	190685868	22.36	5897
1	234808789	247093448	24.48	3789
2	39940529	61792229	21.54	6500
3	36495	10632877	25.72	4288
3	39812713	64231310	22.82	6100
4	140320206	177888785	39.99	7591
5	14343689	26724511	12.58	2499

23ANDME

At 23andMe, a test-taker can use the chromosome browser tool to look at shared segments with any individual with whom he is "sharing genomes," meaning that the two individuals have agreed to share their profiles with each other. In 2016, 23andMe switched to an entirely new user experience with a revamped user interface, new graphics, and revised tools. The chromosome browser for the new site is located at **<you.23andme.com/tools/relatives/dna>**. There, the user can select two people and compare their genomes.

Image P is the first ten chromosomes on the 23andMe chromosome browser, showing the two individuals share a total of two segments. The hatched gray bars, which are difficult to see, represent each chromosome. Each of the purple blocks represents a shared segment of DNA. As with Family Tree DNA's chromosome browser, the segments shown in the 23andMe browser are *not* perfectly to scale in the chromosome browser, so

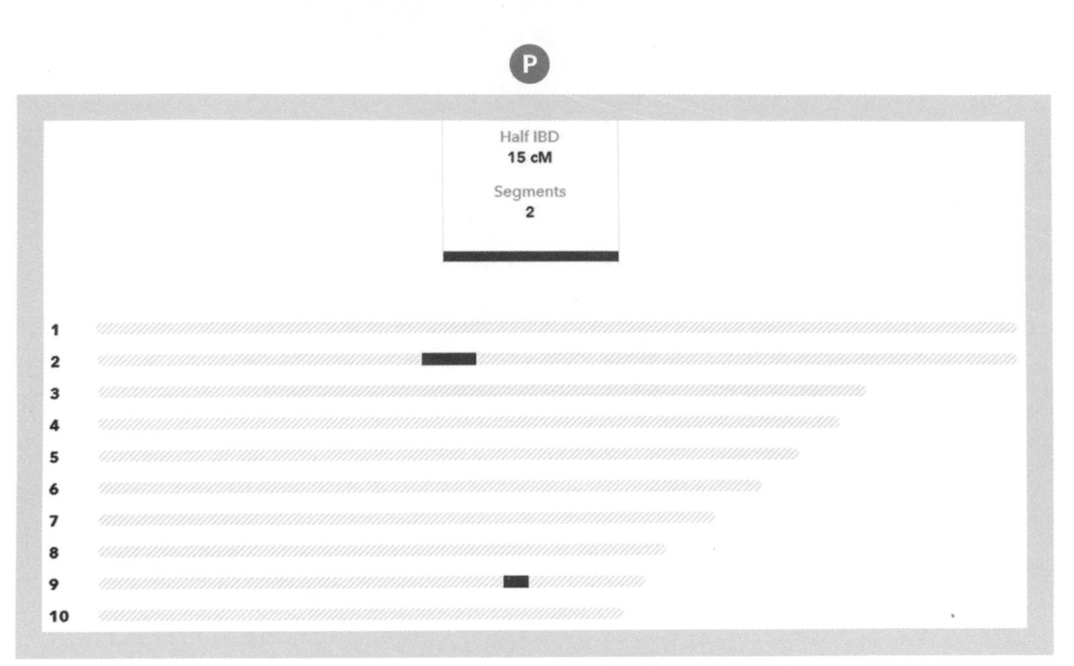

23andMe's chromosome browser also allows for a more detailed comparison between your and your ancestor's DNA.

evaluating the size of segments based on visual appearance alone can be misleading.In addition to showing shared segments on chromosome 1 to 22, 23andMe also shows shared segments on the X chromosome.

If the test-taker clicks on a segment on the chromosome browser, he'll see a pop-up box that shows the chromosome number, the approximate start and stop locations, the total size of that segment, and the number of SNPs tested in the segment. All of the information about shared DNA, including the chromosome number and start and stop location for each segment, can be viewed in a table or downloaded to a spreadsheet.

Analyzing Genetic Cousins Identified by the Testing Company

Each of the three major testing companies, 23andMe, AncestryDNA, and Family Tree DNA, compare the test-taker's DNA to the DNA of every other test-taker in the company's own database. If the two sets of DNA have a segment with the same sequence, and if the length of that segment satisfies the thresholds previously discussed, then the individuals will be identified as genetic cousins or "matches."

Note that while all three companies and all third-party tools use the word "match" to refer to two or more people who are identified as sharing a segment of DNA, the word "match" does not necessarily mean that two people share a recent common ancestor; for example, the two or more people may share that segment by chance or due to a sequencing or interpretation error.

In this section, I'll discuss how to carefully evaluate each company's "matches."

23ANDME

The match list at 23andMe is called DNA Relatives and shows the test-taker's closest genetic relatives listed in order, starting with the individual who shares the most DNA with the test-taker (Image Q).

Unlike AncestryDNA and Family Tree DNA, 23andMe has a default privacy barrier between people in the DNA Relatives list. Because of this privacy barrier, people identified as genetic cousins are not automatically revealed to the test-taker. Instead, the test-taker will only see the identified persons' sex, predicted relationship, mtDNA haplogroup, and the Y-DNA haplogroup (for males). In the profile for anonymous matches, the test-taker will find a Request to Share button that allows users to share ancestry reports *only if* the anonymous match checks for and accepts sharing requests.

The test-taker can also send a message to anonymous matches by clicking on the match and using the message box in the far right column of the profile. Another way to

Q

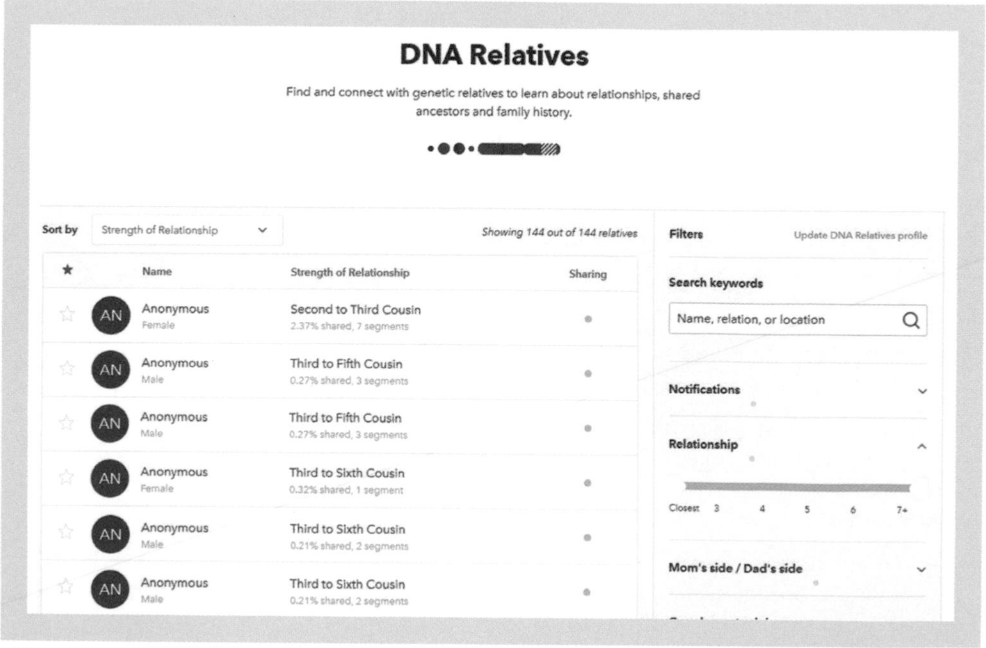

The names of and details about suggested genetic cousins at 23andMe are behind a privacy barrier. In order to see your match's information, that match will need to approve a request to share his ancestry report.

connect with the genetic cousin is sending a personalized communication, such as an online family tree where the match can go for more information.

Given the size of the 23andMe database, most people with European ancestry will have a significant number of genetic matches—most from colonial locations like the United States, Canada, Australia, and New Zealand. Recently, more and more people are also testing from Ireland and the United Kingdom. However, people with mostly Asian and African ancestry may have fewer genetic relatives, since those areas of the world have not experienced widespread genetic testing.

ANCESTRYDNA

The "match list" at AncestryDNA is called DNA Matches and does not have a specific cap. Genetic relatives are listed in order starting with the individual who shares the most DNA with the test-taker. In the example in image R, the test-taker's closest match is a first cousin.

You'll see relevant information for each match including username, relationship range, last login, and information about whether the individual has a family tree linked to his account. Clicking on a username will show the user's profile. If the user has a family tree associated with his DNA test, his genetic relatives will be able to review the family tree to look for surnames and/or places in common.

If the test-taker has a *public* family tree associated with the DNA test results, AncestryDNA will compare that tree to the tree of genetic matches to try to find common ancestors. If a potential common ancestor who is similar enough is identified in the two trees, the genetic relative will have a "shaky leaf hint." Shaky leaf hints are rare, and are generally more successful with larger and more complete family trees. Shared hints should be reviewed as hints, not as proof or evidence of a relationship. The fact that two people share DNA and share a common ancestor does not necessarily mean the shared DNA must have come from that shared ancestor!

Like 23andMe, test-takers with European ancestry will have a significant number of genetic matches identified in the list. However, AncestryDNA is actively advertising in and targeting countries such as Canada, Australia, and the United Kingdom (with more nations to come), and thus an increasing number of people in the database will be from other regions of the world.

FAMILY TREE DNA

The "match list" at Family Tree DNA is called Family Finder – Matches, and, like AncestryDNA, does not have a specific cap. Genetic relatives are listed in order starting with the individual who shares the most DNA with the test-taker. In image S, the test-taker's closest matches are her two grandchildren.

R

1ST COUSIN

Possible range: 1st - 2nd cousins
Confidence: Extremely High
7690 people
VIEW MATCH

Possible range: 1st - 2nd cousins
Confidence: Extremely High
89 people
VIEW MATCH

2ND COUSIN

Possible range: 1st - 2nd cousins
Confidence: Extremely High
34 people
VIEW MATCH

Possible range: 2nd - 3rd cousins
Confidence: Extremely High
No family tree
VIEW MATCH

Ancestry DNA's suggested matches page links directly to individuals' family trees, allowing you to compare and evaluate a potential match's family history to determine if you are, in fact, related. Names of matches have been blurred for privacy.

S

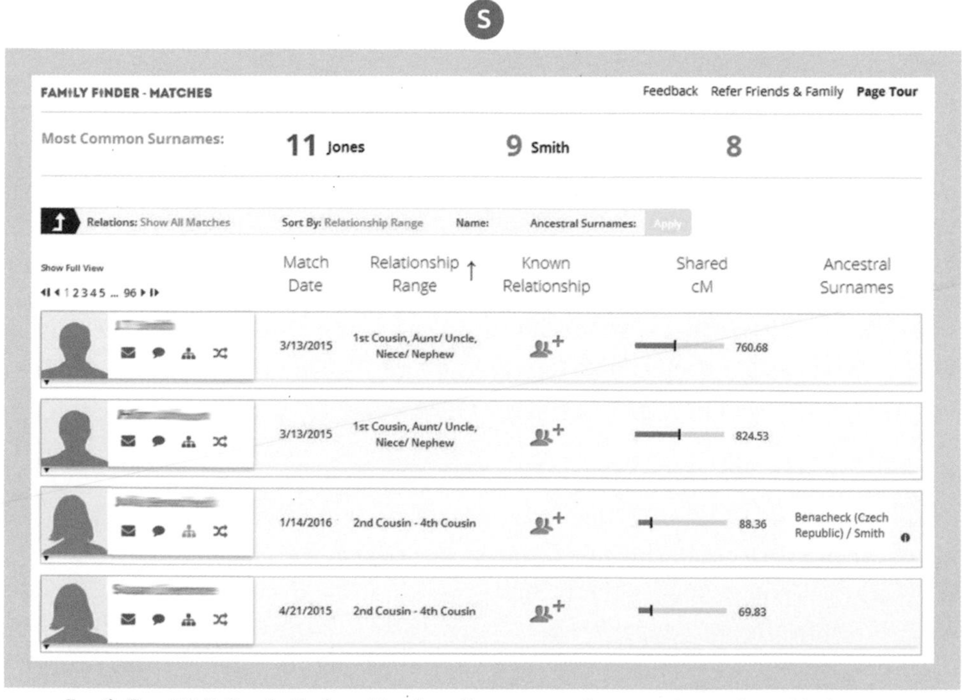

Family Tree DNA's Family Finder—Matches allows you to view genetic matches with the users you share the most DNA with on the top. Names of matches have been blurred for privacy.

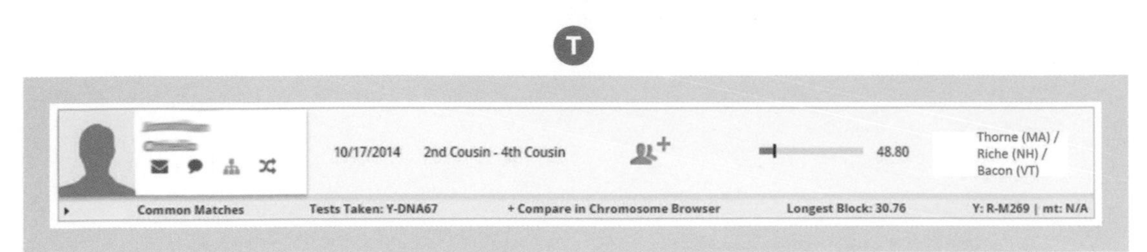

Family Tree DNA provides specific information about matches, including how many shared cMs of DNA, the individual's surnames in his profile (which are bolded if they match yours), and (for male users) a Y-DNA haplogroup. This match's name has been blurred for privacy.

Family Tree DNA provides a wealth of information about each genetic relative, especially if the genetic relative has added certain facts and/or a family tree to the member profile. In image T, the genetic relative is predicted to be a second to fourth cousin and the two share a total of 48.80 cMs, of which the longest segment is 30.76. The user has some surnames in his profile (which will be bolded if any match the surnames listed in the other test-taker's profile) that can quickly be reviewed by hovering over them. Further, this user has taken a Y-DNA test and his Y-DNA haplogroup is R-M269 (his terminal SNP).

Clicking on the username of a genetic relative in the match list will result in a pop-up that provides even more information (if the user has populated these fields), including Y-DNA and mtDNA haplogroup, most distant known paternal and maternal ancestors, and (most importantly for communication and collaboration) an e-mail address so you can contact the match.

Family Tree DNA's database largely comprises test-takers from the United States, Canada, the United Kingdom, and Australia, although (like the other two companies) it includes test-takers from all over the world.

Using atDNA: 'In Common With' Tools

One of the most important tools at AncestryDNA and Family Tree DNA is the In Common With (ICW) tool. ICW tools allow a test-taker to see which of his identified genetic relatives are shared in common with a person in the match list.

AncestryDNA: Shared Matches

AncestryDNA's ICW tool is called Shared Matches and is accessible via the Shared Matches button in a genetic relative's profile (image U).

There is an important limitation of Shared Matches: It will only work for fourth cousins or closer. In other words, for a Shared Match to show up, that Shared Match must be a fourth cousin or closer to *both* the test-taker and the genetic relative with whom the test-taker is checking for shared matches.

U

Member since 2010, last logged in last week

SEND MESSAGE

Predicted relationship: 3rd Cousins
Possible range: 3rd - 4th cousins (What does this mean?)
Confidence: Extremely High

Add note

Ethnicity
Regions: Great Britain, Asia East, Native American, Europe West, Ireland, Polynesia
Trace Regions: Asia Central, Iberian Peninsula, Mali, Caucasus, Africa North, Asia South, Cameroon/Congo, Scandinavia, Europe East, Italy/Greece

PEDIGREE AND SURNAMES | MAP AND LOCATIONS

193 people

VIEW FULL TREE

Ancestry.com's Shared Matches tool allows you to view other users who have DNA matches in common with you, plus gives a predicted relationship with a confidence interval. This user's name has been blurred for privacy.

V

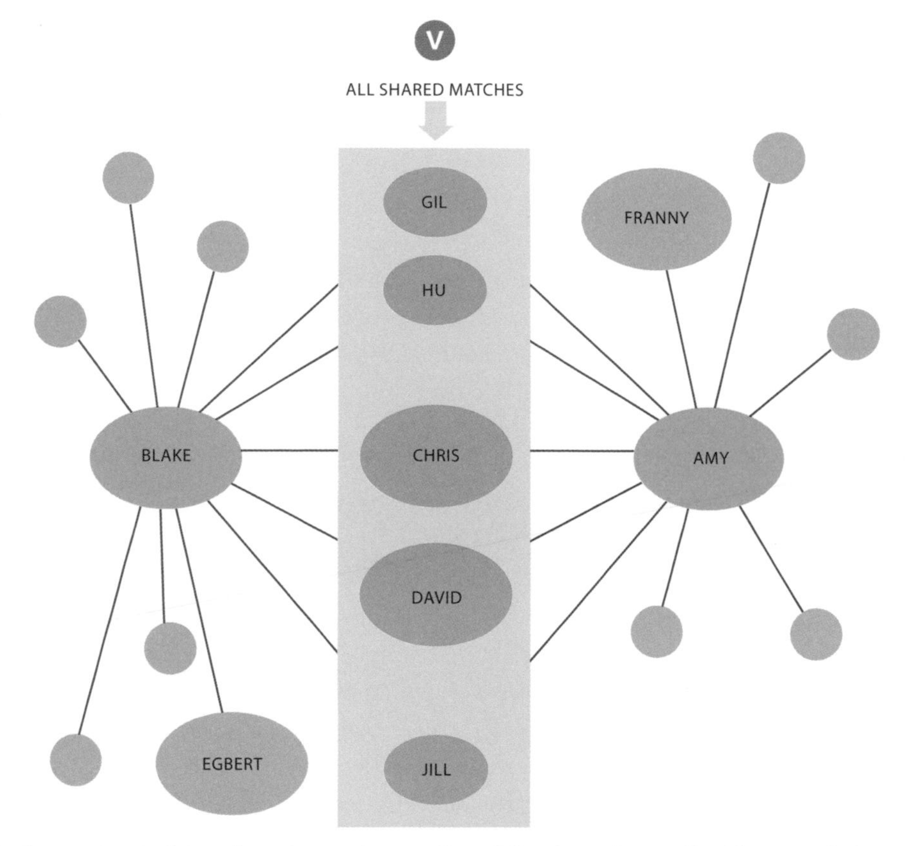

Navigating a network of shared matches can be complicated. Creating a diagram like this one can help you sort out which ancestors you and another test-taker at AncestryDNA do (and do not) have in common.

Let's use an example. Assume the test-taker, Blake, has a great-aunt named Amy who has tested at AncestryDNA. Blake sees Amy in his match list, so he clicks on the View Match button to see Amy's profile. There, he clicks on the Shared Matches button to see the DNA matches he shares with Amy. He obtains a list that includes a third identified genetic relative named Chris. In order for Chris to appear in that list, he must satisfy two criteria: Chris *must* be a fourth cousin or closer to Blake and *must* be a fourth cousin or closer to Amy. If Chris is a distant cousin to Amy, he won't appear in the Shared Matches list *even if* he is a match shared in common with Amy and Blake.

In image Ⓥ, which works off the same example, Blake and Amy both have an array of matches. Some of those matches are shared in common, as shown in the region highlighted in gray. Of those, only Chris and David are fourth cousins or closer to *both* Blake and Amy, so only Chris and David will show up in the Shared Matches list. While (based on other genealogical evidence) Blake and Amy also share Gil, Hu, and Jill in common, they are all more distant matches and won't show up in the Shared Matches list. Further, close matches like Franny and Egbert won't show up because they aren't shared by both Blake and Amy. Obviously, genealogists can only very rarely use the absence of a match from an ICW group as evidence.

Family Tree DNA: The ICW and Matrix Tools

Family Tree DNA offers two ICW tools. The first is the Common Matches tool, which is accessible by clicking the double arrow below an identified genetic match's username. This will provide a list of all individuals in the genetic matches list who are identified as sharing DNA with *both* the test-taker and the individual for which the double arrow was clicked. Unlike AncestryDNA, there is no restriction on the prediction of the genetic relationship, so *all* matches in common with the two compared individuals will be identified on this list. For example, if a mother and daughter have tested and the child does an ICW comparison with her mother, a very large percentage of the daughter's entire match list can be shared with the mother.

The second ICW tool at Family Tree DNA is called the Matrix tool (image Ⓦ). The Matrix tool is accessible by clicking Matrix in the Family Tree DNA Dashboard. The Matrix tool allows a test-taker to select up to ten of her identified genetic relatives and compare their atDNA sharing status in a grid or matrix.

In the following example (image Ⓧ), eight people have been added to the Matrix tool by John, the test-taker. The way in which individuals are grouped in the tool reveals three distinct patterns. First, Art, Bob, and Cary appear in a cluster, meaning that the test-taker John is identified as a genetic match with Art, Bob, and Cary, all of whom are identified as genetic matches with each other. In addition, you can see a second,

W

FAMILY FINDER - MATRIX BETA

Feedback Refer Friends & Family

The **Family Finder Matrix** page allows you to select up to 10 people and compare their Family Finder relationships in a grid (matrix).

The page defaults to two lists:
- Matches: These are Family Finder matches who can be added to the grid.
- Selected Matches: These are Family Finder matches who are currently included in the grid.

Add matches to the matrix by clicking a name or names on the Matches list and then clicking the Add button. Remove matches from the matrix by clicking a name or names in the Selected Matches list and then clicking the Remove button. The grid displays under the list as you begin to add matches to the Selected Matches list. The grid shows those who share a genetic relationship according to Family Finder results with a white check mark on a blue background. When two matches do not match each other, the grid shows a blank white square.

Matches | **Selected Matches**

Add » | « Remove | Move Up | Move Down

The Family Tree DNA Matrix allows users to select matches and compare their shared DNA in a grid or table. Names of matches have been blurred for privacy.

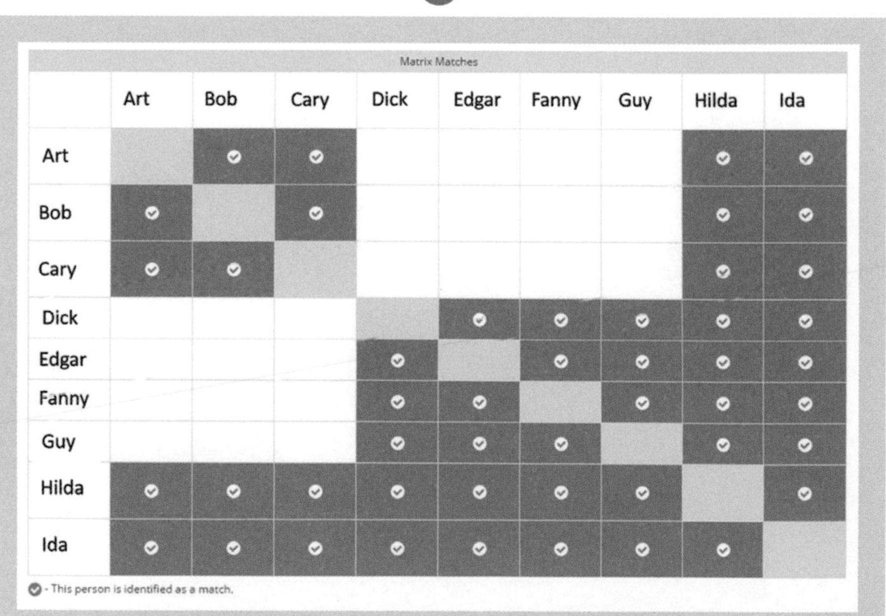

Matrix Matches

	Art	Bob	Cary	Dick	Edgar	Fanny	Guy	Hilda	Ida
Art		✓	✓					✓	✓
Bob	✓		✓					✓	✓
Cary	✓	✓						✓	✓
Dick					✓	✓	✓	✓	✓
Edgar				✓		✓	✓	✓	✓
Fanny				✓	✓		✓	✓	✓
Guy				✓	✓	✓		✓	✓
Hilda	✓	✓	✓	✓	✓	✓	✓		✓
Ida	✓	✓	✓	✓	✓	✓	✓	✓	

✓ - This person is identified as a match.

Test-takers can get a lot from Family Tree DNA's Matrix data, such as this one in which matches who match the both test-taker and each other are indicated in blue. Note how some individuals "cluster" together, suggesting these individuals are related to each other and not necessarily to other matches.

Dick-Edgar-Fanny-Guy group. And third, Hilda and Ida share DNA in common with John, Art, Bob, Cary, Dick, Edgar, Fanny, and Guy. In this particular example, Hilda and Ida are children of John, so it isn't surprising to see them matching everyone in this list.

Note this does *not* mean that all of the individuals in a matrix share a common ancestor with all the other members, as there are scenarios in which a test-taker will share common ancestors with some individuals but not others. For example, if John and Art share a common ancestor while John and Bob and Cary share a *different* common ancestor, Art could happen to share yet another ancestor who isn't related to John with Bob and Cary.

Limitations of ICW Tools

People do not necessarily share a common ancestor just because the ancestor appears in an ICW tool. In the following example (image Y), test-taker George runs an ICW analysis at AncestryDNA and/or Family Tree DNA for an identified genetic match. George discovers that he and Thomas share a third identified match in common: Judy. George is excited and concludes that all three of them share the *same* common ancestor. However, George has jumped to an inaccurate conclusion.

In fact, George and Thomas share Common Ancestor 1, while George and Judy share Common Ancestor 2. But what George can't determine from the ICW tools alone, is

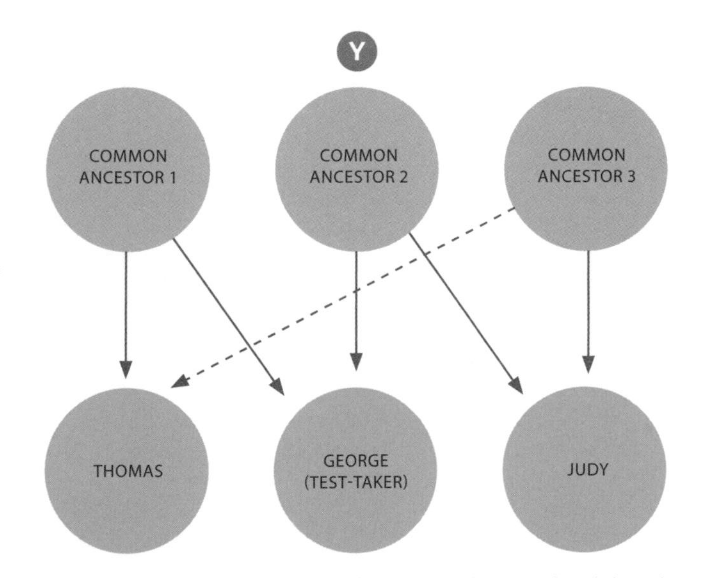

Shared matches can be deceiving. Three individuals can be considered shared matches, but not all three necessarily share the same common ancestor, as in this example.

whether Thomas and Judy share Common Ancestor 1, Common Ancestor 2, or—as is the case here— an entirely different Common Ancestor 3. Note that George does *not* share Common Ancestor 3 with Judy or Thomas. Additional information will be needed to confirm George's hypothesis. For example, if all three of the test-takers share the same segment of DNA in common, this might be stronger evidence for George's conclusion. Alternatively, George can compare the family trees of the three individuals where he might learn of the situation described above.

Other Uses for atDNA Testing

Genealogists use the results of atDNA testing for two main purposes: cousin matching (which we've already discussed in depth) and ethnicity analysis. Ethnicity estimates provided by testing companies attempt to break down the test-taker's DNA into continental or regional sources. Although these estimates are notoriously poor, they can have genealogical applications, particularly if the research question involves a recent ancestor with a distinct ethnicity. For example, finding numerous segments of DNA which the testing company identifies as "African" can support a hypothesis for a recent ancestor with African ancestry. We'll discuss ethnicity estimates in more detail in chapter 9.

In addition to cousin matching and ethnicity analysis, the results of atDNA testing can have other uses. For example, genealogists use cousin matching and family trees to "map" or assign segments of their DNA to ancestors. If test-takers Aaron and Brenda know that they share a segment on chromosome 7 in common, and trace that segment to their great-grandfather, Marshall, they can reasonably assume that segment on chromosome 7 came from their great-grandfather. Thus, when future genetic matches share that segment with them, they know to review their great-grandfather's line in order to find a common ancestor.

Genealogists are beginning to recreate portions of the genomes of ancestors using atDNA. To do this, they test multiple descendants of an ancestor, who are unlikely to have ancestry through lines other than that ancestor. The segments of DNA those descendants share in common, therefore, are predicted to have come from the shared ancestor and can be pieced together to recreate portions of the ancestor's genome. As more descendants are tested, more pieces of the ancestor's DNA can be identified.

Some genealogists are also using atDNA test results to learn about their health and propensity for certain illnesses or conditions. Although the connection between DNA and health is still poorly understood—and what is understood suggests that DNA plays a smaller role in most health conditions than was once predicted—it is possible to analyze a test-taker's DNA for health purposes. 23andMe, for example, provides health information to test-takers as part of its atDNA test. In addition, there are third-party tools that analyze the test-taker's raw DNA data and provide a report of propensities for certain health conditions.

These are just a few of the powerful uses of atDNA, and there will be more as additional people take atDNA tests and as new tools are developed by testing companies and independent programmers.

CORE CONCEPTS: AUTOSOMAL-DNA (ATDNA) TESTING

- **Autosomal DNA (atDNA) refers to the twenty-two pairs of chromosomes, called autosomes, in the nucleus of the cell.**

- **A child inherits 50 percent of atDNA from the father and 50 percent of atDNA from the mother.**

- **atDNA testing is done by analyzing hundreds of thousands of SNPs throughout the twenty-two pairs of chromosomes.**

- **atDNA test results are used to fish for genetic cousins by approximating how many generations have passed since two matches shared a common ancestor. The closer the relationship, the better the estimate.**

- **Not all genealogical cousins will share atDNA. Cousins and relatives at the second-cousin or closer level are always expected to share DNA. Beyond a second-cousin relationship, the likelihood of sharing DNA with a cousin decreases rapidly.**

- **Each of the atDNA testing companies—23andMe, AncestryDNA, and Family Tree DNA—uses atDNA test results to estimate ethnicity and find genetic cousins. Each of the companies offers tools to analyze the results of testing and connect with genetic cousins.**

- **atDNA test results can be very useful for examining specific genealogical questions, such as whether two people share a recent common ancestor.**

DNA in Action

What Is the Relationship?

Genealogist Allen, age twenty-five, has tested his atDNA at all three testing companies. He periodically logs into his accounts to check for new matches, and when he logs into Family Tree DNA, he discovers a new close match with the username "NYgreen3." This match shares 1025 cMs with Allen, and is predicted to be a "1st cousin, half-sibling, grandparent/grandchild, aunt/uncle, or niece/nephew." Allen doesn't recognize the username or the e-mail address associated with the account, and no other information is provided. The shared matches feature reveals that NYgreen3 matches Allen's maternal relatives, particularly those on his maternal grandmother's line.

To figure out how NYgreen3 might be related to him, Allen turns to the Autosomal DNA Statistics page of the ISOGG wiki **<www.isogg.org/wiki/Autosomal_DNA_statistics>**, which provides a table of the expected amount of DNA shared between people having certain genealogical relationships. The relevant rows of the table are provided below:

Percentage	cMs Shared	Relationship
25%	1700	Grandparent, uncle/aunt/niece/nephew, half-sibling
12.5%	850	Great-grandparent, first cousin, great-uncle/aunt/niece/nephew, half-uncle/aunt/niece/nephew
6.25%	425	First cousin once removed, half first cousin

According to this table and the 1025 cMs shared by the genetic matches, NYgreen3 is closest to the 12.5 percent of sharing and thus is predicted to be a great-grandparent, first cousin, great aunt/uncle/niece/nephew, or half-aunt/uncle/niece/nephew. Without more information, however, Allen is unable to determine the exact relationship.

Allen contacts the individual and learns that NYgreen3 is male, seventy-five years old, and adopted. The fifty-year difference between Allen and NYgreen3 (whose real name is Joseph) suggests that he is not a first cousin or a half-uncle/nephew. He is also unlikely to be Allen's great-grandfather, as he was in a different country when Allen's maternal grandparents were conceived. This suggests, therefore, that NYgreen3 (Joseph) Is potentially Allen's great-uncle, his maternal grandmother's brother. Indeed, additional research shows that Joseph was raised near the town were Allen's grandmother was born, shedding light on both Allen's and Joseph's family trees and potentially fostering a new meaningful family connection.

DNA in Action

Was She Native American?

Like many other families in the United States, particularly those with colonial ancestry, the Cornwall family has a long-standing oral tradition of a Native American ancestor. Andrea Cornwall is interested in genealogy and asks her paternal grandfather Caleb Cornwall about this ancestor. He tells her that according to family tradition, the Native American ancestor saved his grandfather Cornwall from death and then married him, and together they had two children.

Andrea would like to confirm—or reject—this story using DNA testing. She does a little research and learns that her great-great-grandmother, the Native American according to family legend, was named Abigail and died young during childbirth while giving birth to Caleb's father.

Unfortunately, because this ancestor is neither a direct Y-DNA or mtDNA ancestor, Andrea can only do an atDNA test of her grandfather Caleb, her mother Susan (Caleb's daughter), or herself. Since Caleb will have more of his grandmother's atDNA, Andrea asks him to take an at DNA test. If Abigail was indeed a Native American as family legend reports, then Caleb's DNA should report a significant percentage of Native American DNA (as much as 25 percent, potentially, since about 25 percent of his DNA will have come from Abigail).

When the test results come back from the testing company, Caleb receives the following ethnicity estimate together with his match list:

Ethnicity	Percentage
African	0%
Asian	0%
European	97.5%
Native American	2.5%

Based on the results, it is unlikely that Andrea was Native American herself, since Caleb's percentage of Native American ancestry is very low. She may have had Native American ancestors, but additional testing will be necessary to examine this possibility.

7

X-Chromosomal (X-DNA) Testing

What does it mean to share X-chromosomal DNA (X-DNA) with a match? One of the most powerful advantages of Y-chromosomal DNA (Y-DNA) and mitochondrial DNA (mtDNA) is that you always know *exactly* what ancestor in the family tree provided that piece of DNA. In contrast, with autosomal DNA (atDNA), *any* of your ancestors could have provided a segment of DNA. X-DNA falls between these two extremes; while there are many ancestors who could have contributed to your X-DNA, they make up only a small subset of your entire genealogical family tree. Thus, sharing X-DNA with a match means you only have to search that subset of your tree for the common ancestor. In this chapter, we'll learn about X-DNA and how it can be utilized to explore common ancestry with your genetic matches.

The X Chromosome

The **X chromosome** (image A) is one of the twenty-three pairs of chromosomes found in the nucleus of the cell, and is one of the two sex chromosomes, the other being the Y chromosome (which, as you'll recall, is found only in males). Unlike Y-DNA, both men and women have X-DNA. Women have two X chromosomes, one inherited unchanged

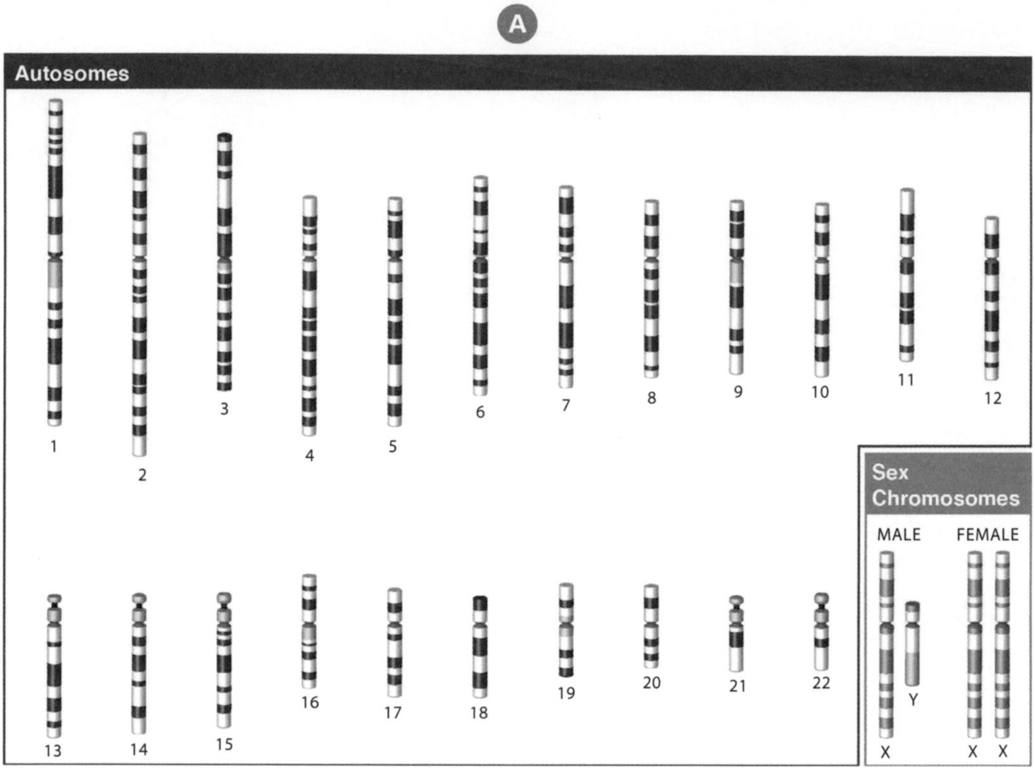

The X chromosome, together with the Y chromosome, makes up the sex chromosomes, which can provide valuable information for genealogists. Note that only one copy of each of autosomal chromosome is shown above; in reality, each person has two copies of each of the twenty-two autosomes. Courtesy Darryl Leja, National Human Genome Research Institute.

from their father and one inherited from their mother. Men have just one X chromosome that they inherited from their mother.

The X chromosome is a relatively large chromosome of approximately 150 million base pairs, and contains about two thousand of the estimated twenty to twenty-five thousand genes found throughout the entire human genome.

The Unique Inheritance of X-DNA

Like both mtDNA and Y-DNA, X-DNA has a unique inheritance pattern that makes it valuable for genetic genealogy testing. A mother *always* passes down an X chromosome to *all* of her children, either male or female. In contrast, a father will only pass down his X chromosome to his female children. As a result, a father and son always break the transmission of X-DNA in a family tree.

A woman has two X chromosomes: one copy she received from her mother and one copy she received from her father. If a woman has children, she will pass down an X chromosome, although this inheritance can results in a few different scenarios based on random events during an egg cell's creation. Sometimes a mother will pass down a full X chromosome to her child completely unchanged from the copy she received from either her mother or her father. In this scenario, the child will share X-DNA with only one maternal grandparent. Other times, the mother will jumble or recombine her two copies of the X chromosome, and the copy she passes down to her son or daughter will be a mixture of the two. In this scenario, the child will share at least some X-DNA with both maternal grandparents. These scenarios are equally possible.

A father, however, always passes down the X chromosome without recombination. Although the tips of the Y chromosome and the X chromosome will sometimes recombine,

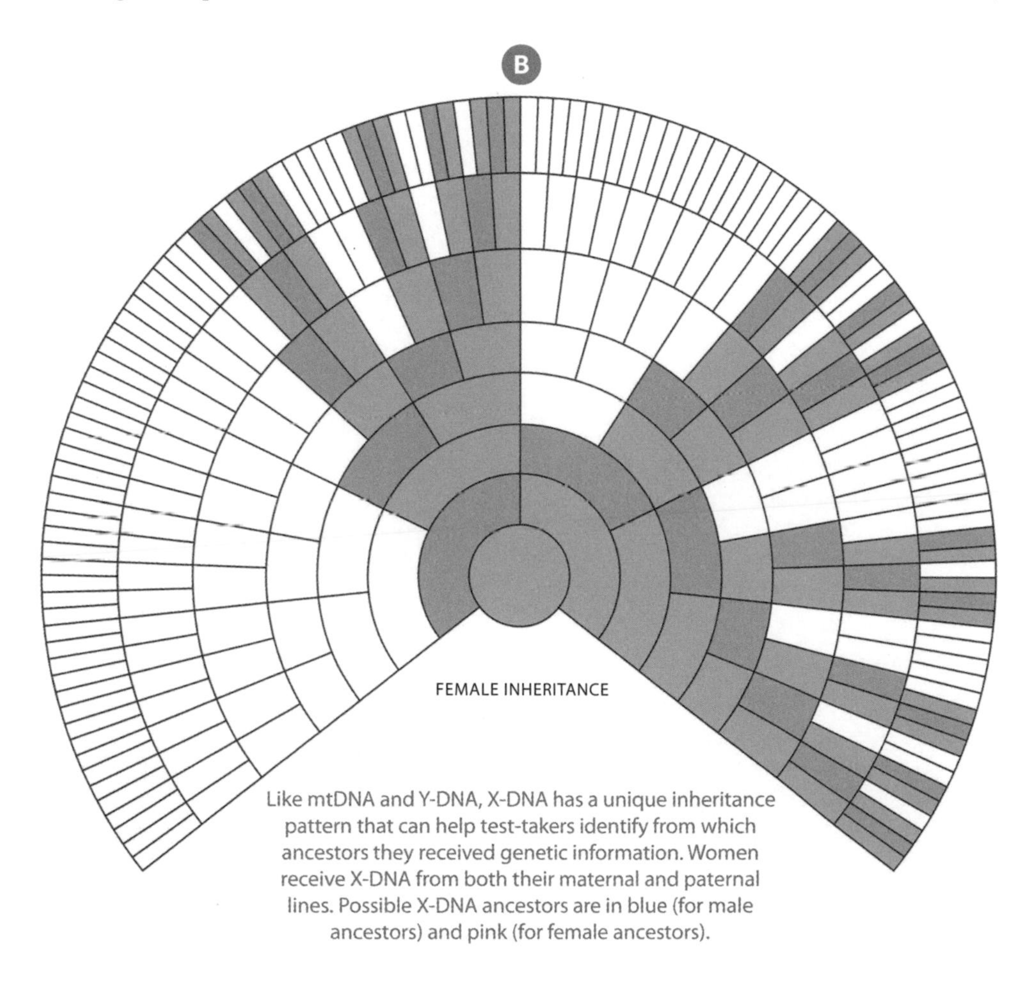

Like mtDNA and Y-DNA, X-DNA has a unique inheritance pattern that can help test-takers identify from which ancestors they received genetic information. Women receive X-DNA from both their maternal and paternal lines. Possible X-DNA ancestors are in blue (for male ancestors) and pink (for female ancestors).

these regions of the Y chromosome are not utilized for genetic matching. Accordingly, the child will share X-DNA only with the paternal grandmother. A child will only share X-DNA with a paternal grandfather indirectly, through other lines of the family tree.

Image B shows the *possible* sources of X-DNA within a family tree for a woman. This tree traces back the possible path of a female X-DNA through seven generations, or to fifth great-grandparents. At that generation, an individual has 128 ancestors (or fewer, if there are recent cousin marriages). Of those 128 ancestors, a woman will have thirty-four potential contributors (thirteen males and twenty-one females) to her two X chromosomes. Since this is a chart for a woman who inherited X-DNA from her mother and father, there are possible sources of X-DNA on both sides of her family tree: The male

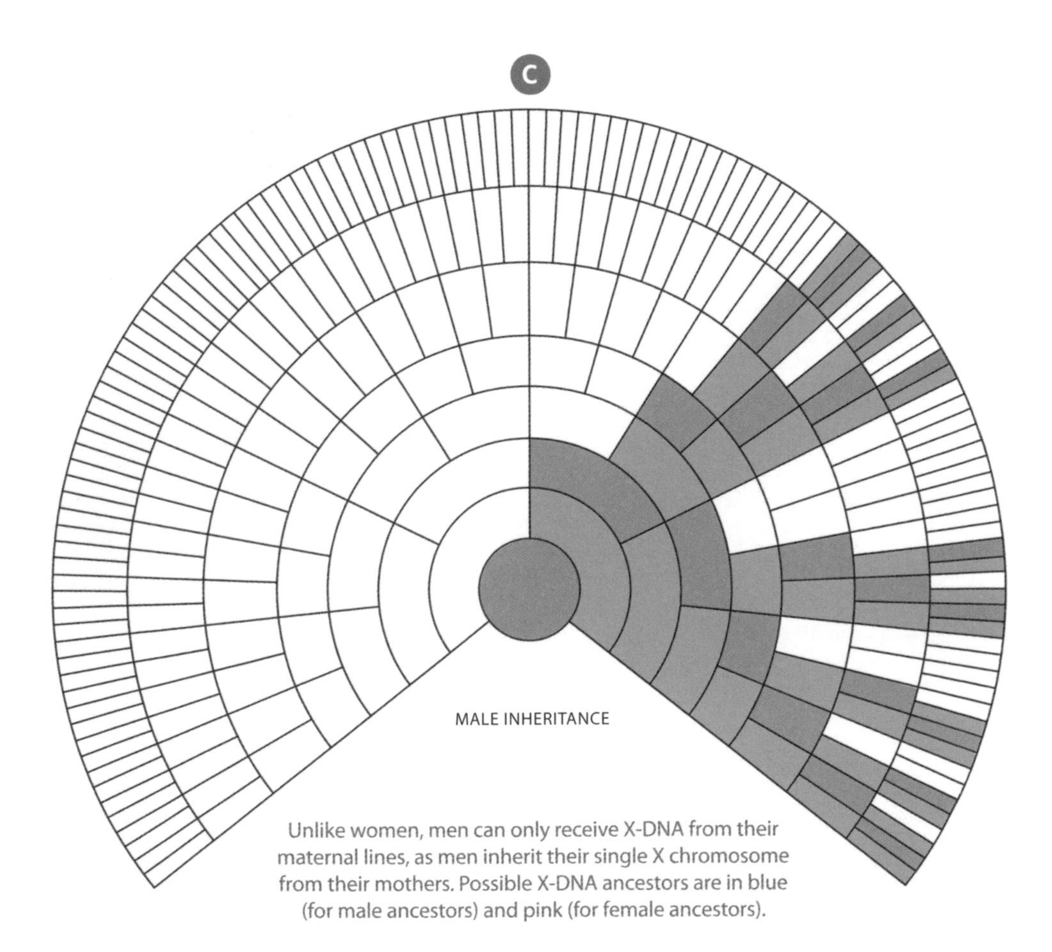

Unlike women, men can only receive X-DNA from their maternal lines, as men inherit their single X chromosome from their mothers. Possible X-DNA ancestors are in blue (for male ancestors) and pink (for female ancestors).

possible sources of X-DNA are highlighted in blue, and the female possible sources of X-DNA are highlighted in pink.

Note that although this chart shows the *possible* sources of X-DNA within a family tree for a woman, the actual sources of the woman's X-DNA will be a small subset of the highlighted cells. For example, if the woman inherited her maternal grandfather's X chromosome from her mother, none of her maternal grandmother's family provided X-DNA.

Image C shows the *possible* sources of X-DNA within a family tree for a man. The male possible sources of X-DNA are highlighted in blue, and the female possible sources of X-DNA are highlighted in pink. Since this is a chart for a man who inherited his X chromosome entirely from his mother, only his mother's ancestors could have provided X-DNA. For example, of the 128 ancestors at the seventh generation, only twenty-one of them (eight males and thirteen females) can potentially provide X-DNA to the man. As with the previous chart, the actual sources of the man's X-DNA will be a small subset of the highlighted cells.

As with any isolated autosomal chromosome, the fact that a woman can pass down the X chromosome with or without recombination means that X-DNA sharing with the previous generations can take many different forms. Image D demonstrates X-DNA inheritance through three generations of a family in which the X chromosome either did or did not recombine before it was passed down to the next generation.

Following the X-DNA through this chart to the four grandchildren raises several interesting observations regarding X-DNA inheritance:

1. The paternal grandfather, David, has no daughters in this chart, and thus his X-DNA (indicated in blue) did not pass down to anyone else in this three-generation tree.
2. The maternal grandfather, Nathan, has just a single copy of the X chromosome, and thus he passed down that single copy (indicated in red) completely unchanged to his daughter Susan.
3. The paternal grandmother, Justine, passed down one copy of her X chromosomes without recombination (indicated in green). Benji, therefore, received a full chromosome from either his maternal grandfather or his maternal grandmother (i.e., from one of Justine's parents).
4. The maternal grandmother, Cara, recombined her two copies of the X chromosome when she passed a copy down to her daughter, Susan. Susan, therefore, has X-DNA from three of her four grandparents (Nathan's mother and Cara's two parents).
5. Benji has just a single X chromosome, and thus he passed down that single copy completely unchanged to his two daughters, Ann and Donna.

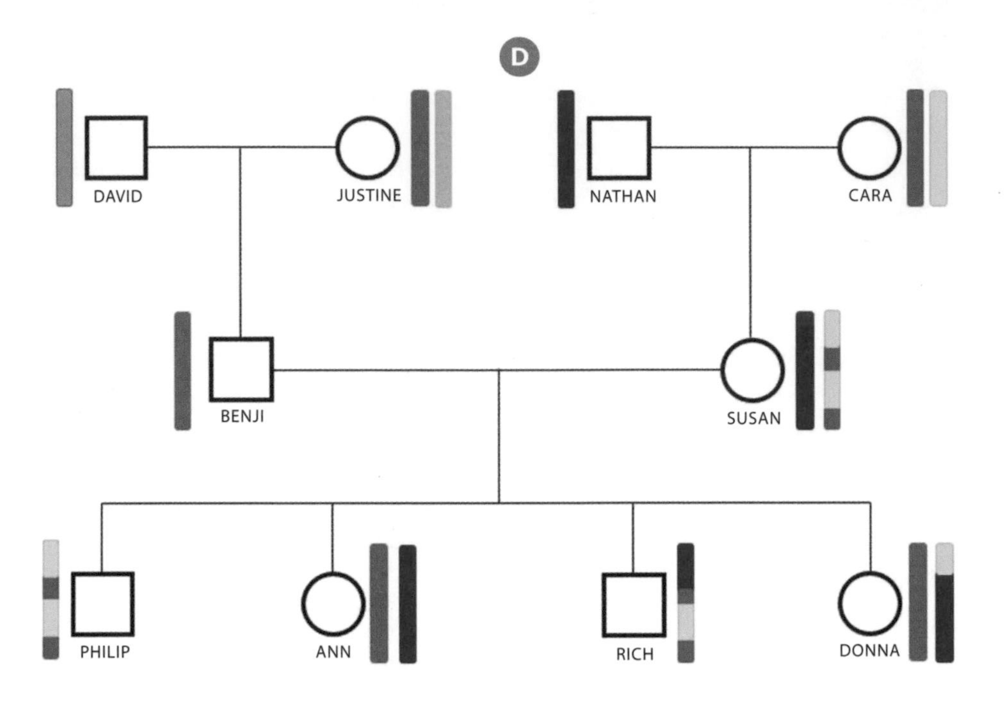

Recombination (in addition to X-DNA inheritance patterns) can drastically affect which X-DNA is inherited through generations. Solid colors represent X-DNA that has not been recombined and so was passed down to the next generation unchanged. Note that males have only one X chromosome while females have two X's.

6. Siblings Philip and Ann each received an X chromosome from their mother without recombination, while siblings Rich and Donna each received a recombined X chromosome from their mother.
7. Siblings Ann and Donna share a full X chromosome in common. This will always be the case for (full) sisters since they always receive the same X chromosome from their father.
8. Philip shares X-DNA with Rich (the blue and purple in the "bottom half" of the chromosome) and Donna (the blue at the top), but none with Ann. It is not uncommon for siblings (who aren't full sisters) to share no X-DNA in common.

How the Test Works

Currently, X-DNA is tested as part of an atDNA test, not as its own test. The test includes between approximately seventeen thousand to twenty thousand single nucleotide polymorphisms (SNPs) on the X chromosome, which will be included in the raw data.

The three main testing companies each treat X-DNA a little differently. Although AncestryDNA <**www.dna.ancestry.com**> tests the X chromosome, it does not use X-DNA when comparing individuals to the database. As a result, you will not have any matches at AncestryDNA that only share X-DNA.

At 23andMe <**www.23andme.com**>, the test-taker's X-DNA is compared to that of other people in the database, meaning that some matches at 23andMe will only share X-DNA. Due to the fact that men have one X chromosome and women have two X chromosomes, the thresholds at 23andMe for comparing men and women will vary. The thresholds for X-DNA can be found in the following table.

Person #1	Person #2	Centimorgan Threshold	SNP Threshold
Male	Male	1	200
Male	Female	6	600
Female	Female	6 (Half-IBD)	1200 (Half-IBD)
Female	Female	5 (Full-IBD)	500 (Full-IBD)

In the table, "Half-IBD," or "half identical by descent," for X-DNA comparisons means that two women share DNA on just one copy of their X chromosomes. Likewise, "Full-IBD," or "full(y) identical by descent," means that the two women share DNA at the same location on both copies of their X chromosomes. The matching threshold for Full-IBD is significantly lower than for Half-IBD. Since only females have two X chromosomes, only females can have half-IBD or full-IBD segments.

At Family Tree DNA, X-DNA matching is reported only if the matches also share atDNA above the matching threshold. Accordingly, you will not have matches at Family Tree DNA that only share X-DNA. As shown in the following table, the matching threshold for X-DNA is significantly lower than the matching threshold for atDNA.

DNA Type	cM Threshold	SNP Threshold
atDNA	7.7	500
X-DNA	1	500

Both Family DNA and 23andMe will show X-DNA matching in their respective chromosome browsers. Image E is a screenshot of the Family Tree DNA chromosome browser that compares the X chromosome of a woman to those of three of her siblings: a sister (orange), a brother (blue), and another brother (green). As the viewer reports, the test-taker shares variable amounts of her X-DNA with each of her siblings.

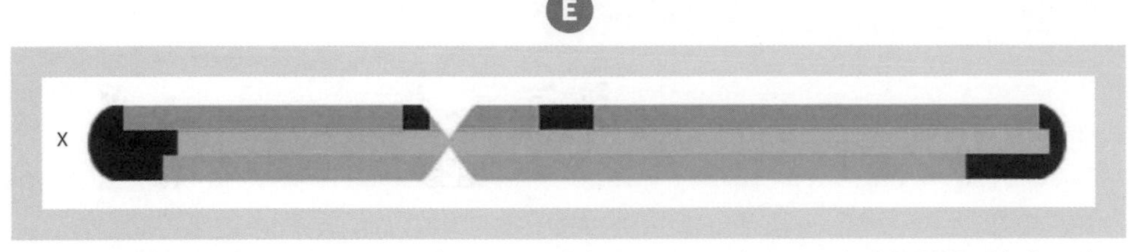

Family Tree DNA has a chromosome browser tool that compares the test-taker's X-DNA with that of other test-taker's. In this case, the tool highlights the X-DNA that the test-takers shared with three other test-taker's: her sister (orange), her brother (blue), and another brother (green).

Limitations of X-DNA Testing and Matching

Genetic genealogists have found that X-DNA matching is not perfect, and can be problematic for several reasons.

By its inheritance pattern, X-DNA can make it difficult to distinguish genetic relationships between two people or predict how much X-DNA two relatives will share. For example, as discussed earlier, the test-taker should share an entire X chromosome with her sister (indicated in orange) but, due to several limitations discussed later, doesn't share certain pieces of X-DNA with her.

Image F further demonstrates this particular limitation of X-DNA. The Family Tree DNA chromosome browser at the bottom compares the X chromosome of a great-grandmother, Alberta, to that of her two male great-grandchildren, Donald and Damian (orange and blue, respectively). Alberta passed down an X chromosome to her son, Bert, and he passed it down—unchanged—to his daughter, Catherine. Catherine then passed down an X chromosome to each of her sons, Donald and Damian. Due to the randomness of recombination, Donald and Damian could have received some, all, or none of Alberta's X-DNA.

The Family Tree DNA chromosome browser can shed some light on this, as it indicates both Donald and Damian received some of Alberta's X-DNA, with one (in blue) receiving significantly more than the other (in orange). Note that because Donald and Damian could have only inherited X-DNA from the maternal grandfather (Bert) or the maternal grandmother (Catherine's mother), the regions they don't share in this chromosome browser view should match X-DNA from their maternal grandmother.

In addition, it is believed that the density of the SNPs tested on the X chromosome is much lower than on comparable chromosomes. The X chromosome is a relatively large chromosome of approximately 150 million base pairs, comparable to chromosome 7 (159 million base pairs). However, the number of chromosome-7 SNPs tested by the three testing companies is nearly double the number of SNPs tested on the X chromosome. As a result, a segment of X-DNA may have relatively few tested SNPs.

With a lower SNP density, there is a greater chance for a segment of DNA to appear like it is a shared segment when in fact it is not a true matching segment. For example, image G compares the two males' X-DNA. If the highlighted SNPs were the only SNPs tested, the two strands of X-DNA would appear to match. However, if the SNP density were increased, results would immediately show that this is not a matching segment. Note this potential hazard is more likely to affect smaller segments of DNA samples, as larger segmented samples will have more SNPs tested.

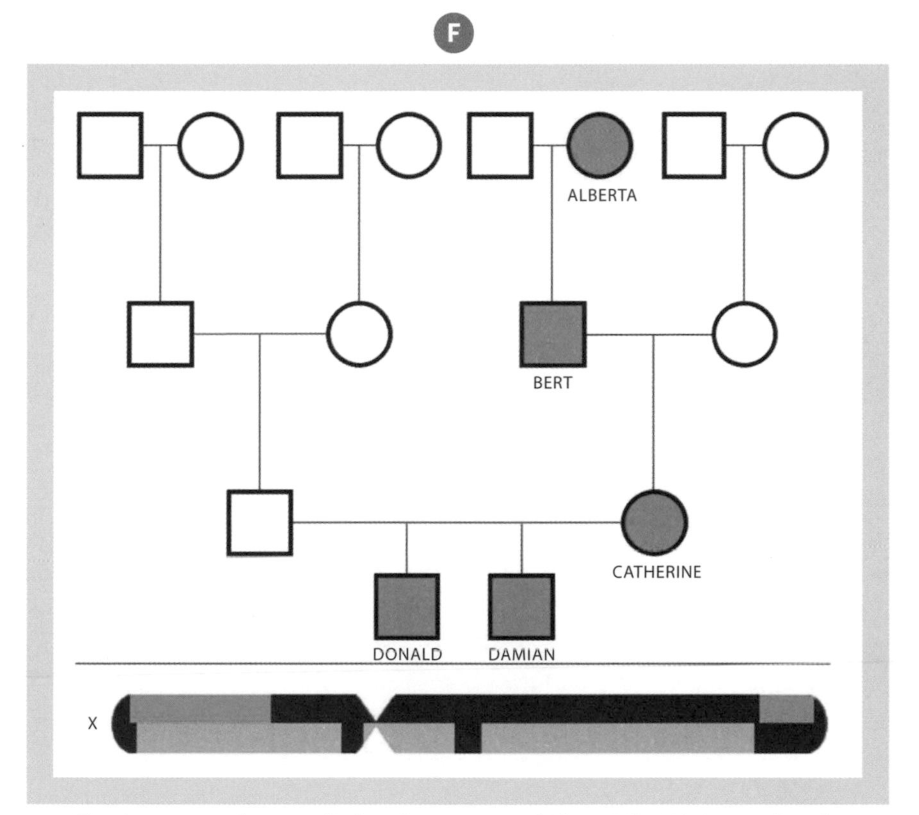

The chromosome browser displays the segments of Alberta's X-DNA that are shared with the X-DNA of her two great-grandsons, Donald (orange) and Damian (blue).

G

-ATCGGCTTAGCAATCATACGTACTCGA-
-GTCAGTTTCCAGCTAAGCATCAGGGGC-

In this example, all the SNPs that an X-DNA test sampled (highlighted in yellow) happened to match. As a result, the X-DNA test would report these two individuals' strands of DNA as matches even though they contain several nonmatching SNPs.

As a result of the current limitations of X-DNA, test-takers should only analyze sufficiently long X-DNA segments. A commonly recommended threshold, for example, is 10 cMs, although some genetic genealogists set even higher thresholds at 15 or 20 cMs. While X-DNA matches absolutely share smaller segments, a genetic genealogist analyzing these small segments does not have enough information to decipher between a true match and a false positive match.

Another limitation of X-DNA matching is the low thresholds used to compare two people's X-DNA. For example, both 23andMe and Family Tree DNA use X-DNA thresholds that are lower than the thresholds for atDNA. At 23andMe, for example, the threshold for comparing the X-DNA of two males is just 1 cM and two hundred SNPs. At Family Tree DNA, the threshold for comparing the X-DNA of any two individuals is just 1 cM and five hundred SNPs. Many genetic genealogists have found this low threshold leads to X-DNA matching that does not appear to be true matching.

Applying X-DNA Test Results in Genealogical Research

Despite its limitations, X-DNA matching can be very useful for genealogy, especially when combined with other types of DNA. For example, sharing both X-DNA and atDNA with a cousin suggests which lines of the genealogical family tree to look for a common ancestor.

However, sharing X-DNA and atDNA with a match suggests—but does not prove—that the atDNA common ancestor is also an X-DNA ancestor. This rule seems counterintuitive at first. After all, if we share both X-DNA and atDNA with a match, doesn't that mean our common ancestor is on one of the X-DNA lines based on the charts we saw earlier in the chapter? Unfortunately, DNA is never that easy! Instead, even though we share atDNA and X-DNA with a genetic match, those segments of DNA could have come from different ancestors. Often, the matching atDNA and X-DNA will come from the same common ancestor. However, the genetic matches will share at least two different common ancestors on different lines just as often, with one line providing the matching atDNA and the other line providing the matching X-DNA (image H).

In addition to multiple ancestors, a genetic match might only share a very small segment of X-DNA that turns out to be a false segment. In this scenario, the genetic matches may spend a considerable amount of time looking for a common X-DNA ancestor who doesn't exist. Deciphering between these possibilities will require an in-depth analysis of both test-taker's family trees, and careful consideration of the size of the X-DNA segments involved.

In addition to the fact that an X-DNA match does not guarantee an atDNA match, genealogists should bear in mind that a lack of X-DNA sharing is almost never informative about a particular relationship. Failing to share X-DNA with another person is almost never evidence of the existence or non-existence of a relationship. There are only a few rare exceptions when two people must share X-DNA: a mother and her children (both male and female), a father and his daughters (who will be full matches with the paternal grandmother), and "full" sisters who have the same father.

Other than these relationships, it is possible that two people who are either closely or distantly related may or may not share X-DNA. For example, while sisters who have the same father will always share a full chromosome, siblings who don't share a father may not share X-DNA. Similarly, brothers and sisters may or may not share X-DNA with their mother. Of course, not sharing X-DNA does not mean that siblings are not related like they thought they were. Instead, they may have received entirely different X-DNA from their mother.

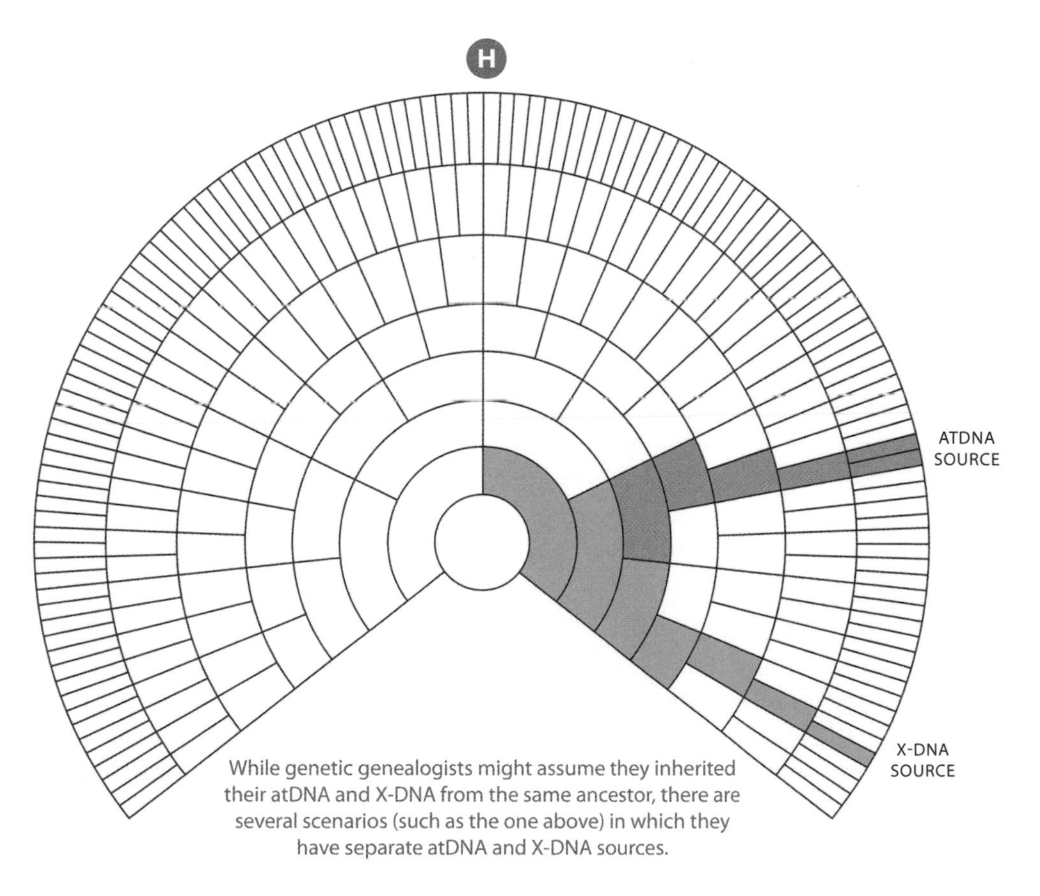

While genetic genealogists might assume they inherited their atDNA and X-DNA from the same ancestor, there are several scenarios (such as the one above) in which they have separate atDNA and X-DNA sources.

I

X

The results from an X-DNA test, such as these for Julia and April, can (and should) be used in conjunction with other pieces of genealogical information.

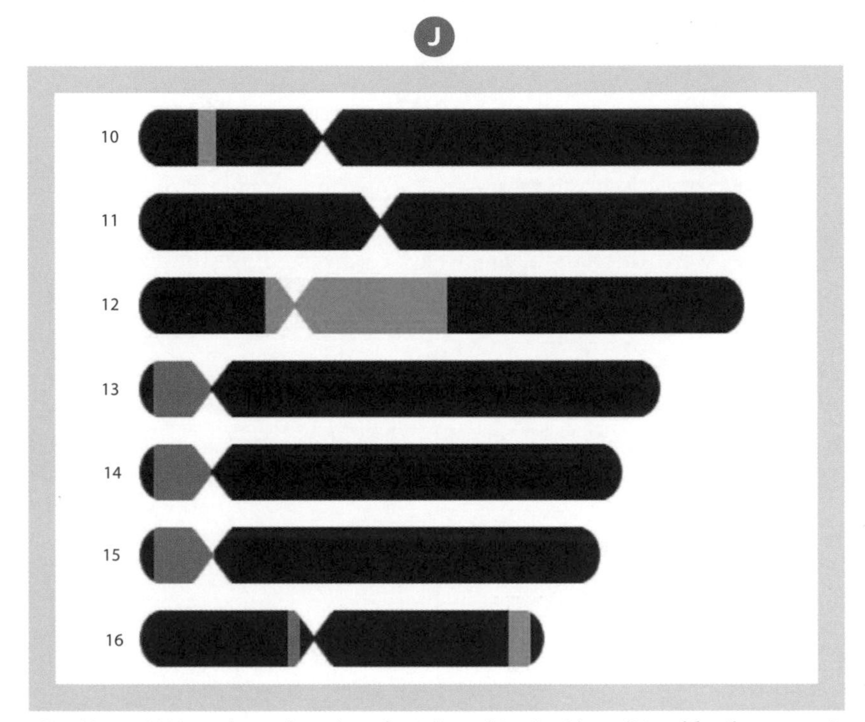

Combine atDNA results, such as these for Julia and April, with traditional family trees and the results of an X-DNA test to draw conclusions about the test-takers' relationships to each other and to a common ancestor. (Note: Gray indicates areas not covered by the test.)

Keeping in mind the limitations and rules outlined in this section, genealogists can analyze an X-DNA match to find the common ancestor or ancestors. X-DNA testing (and analyzing X-DNA and atDNA results with X-DNA inheritance rules in mind) can help shed light on who two individuals' common ancestor might be.

Let's walk through an example in action. In image I, two people share a segment of DNA on the X chromosome (indicated in orange) of approximately 25.28 cMs. Based on the results of an atDNA test (image J), the two individuals also share several segments of atDNA, including segments on chromosome 10 (10 cMs), chromosome 12 (36.94 cMs), and chromosome 16 (16.26 cMs). Family Tree DNA predicts these two, Julia and April,

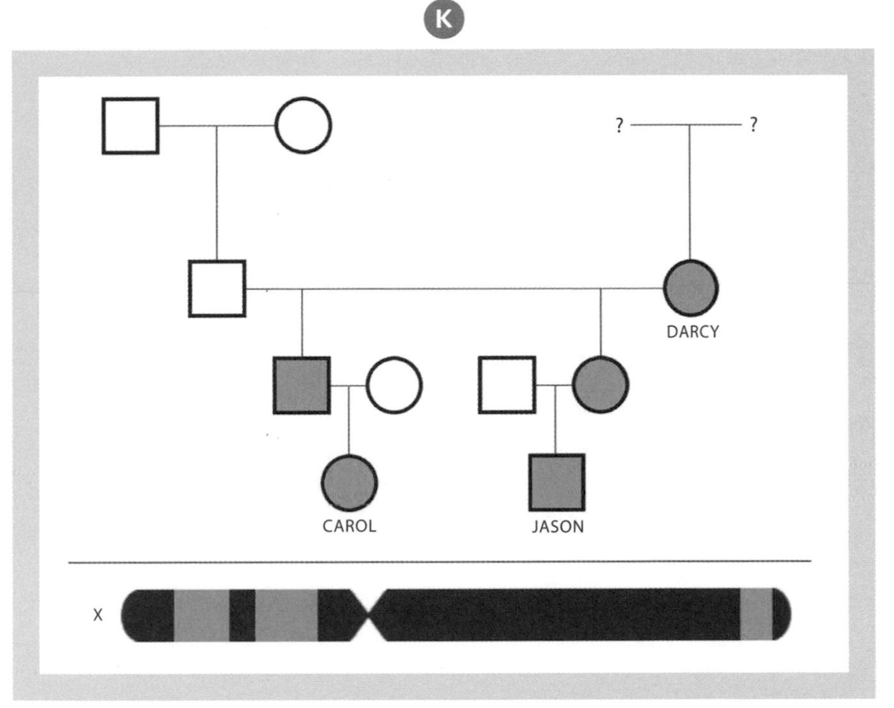

Carol and Jason share the X-DNA highlighted in orange on the chromosome browser. Based on the large amounts of shared X-DNA, they can safely assume they both inherited their X-DNA from their grandmother, Darcy. Carol and Jason can find more of Darcy's relatives and work backwards to find her ancestors by finding individuals with similar X-DNA to theirs.

to be second to fourth cousins. And when Julia and April compare family trees, they discover a potential ancestor in common (a man named Hiram Alden) who would make them fourth cousins. Hiram Alden is an ancestor of Julia's paternal grandfather and an ancestor of April's maternal grandmother.

So is Hiram Alden the common ancestor? X-DNA inheritance patterns tell us no. As shown in the X-DNA inheritance charts earlier in this chapter, Julia could not inherit any X-DNA from her paternal grandfather. Accordingly, while Julia and April may have inherited some segments from Hiram, he cannot be the source of that shared X-DNA. Julia and April must share another ancestor somewhere along their X-DNA lines.

In the next example, shown in image K, Darcy was adopted, and her descendants have no clues about her biological heritage. Darcy and her children are deceased, but two of her grandchildren—Carol and Jason—are living and have both taken an atDNA test that includes X-DNA. When they compare their test results, they see that they share three large segments on the X chromosome (21.65 cMs, 26.83 cMs, and 18.57 cMs). Carol and Jason are curious about where this X-DNA came from and how they can use it to learn about their grandmother's ancestry.

While X-DNA can't provide any definitive answers in this case, it can give Carol and Jason some new avenues of research. The large amounts of shared DNA (indicated in orange) suggest the two share a recent common X-DNA ancestor, and based on this info and their family tree, Carol and Jason likely obtained their shared X-DNA from their grandmother. Carol and Jason could now look for other people who share these segments of X-DNA to find Darcy's other relatives.

CORE CONCEPTS: X-CHROMOSOMAL (X-DNA) TESTING

- **The X chromosome is one of the two sex chromosomes, of which men have one copy (from their mother) and women have two copies (one from their mother and one from their father).**
- **X-DNA is inherited from a small subset of ancestors, meaning that the possible pool of ancestors with whom a test-taker shares an X-DNA cousin is smaller than the pool for the other chromosomes.**
- **X-DNA testing is typically done by SNP testing and only as part of an atDNA test (rather than as a standalone test).**
- **The results of an X-DNA test can be used to fish for genetic cousins.**
- **Due to the low SNP density of current X-DNA tests, as well as the low thresholds utilized by the companies, X-DNA matches must be very carefully scrutinized, and only large X-DNA matching should be pursued.**
- **Sharing X-DNA and atDNA with a match suggests that the atDNA common ancestor is an X-DNA ancestor, but it is also possible that the atDNA and the X-DNA came from different ancestors.**
- **Lack of shared X-DNA is rarely informative about a particular relationship, since there are only a few relationships in which relatives *must* share X-DNA.**

PART THREE

Analyzing and Applying Test Results

8

Third-Party Autosomal-DNA Tools

You've tested with one or more of the testing companies, you've reviewed your ethnicity estimate, and you've gone through your match list. Now what should you do? How do you maximize your testing dollars to wring every piece of useful information out of your DNA test(s)? Third-party tools—both free and for a fee—provide new tools and avenues of research for genealogists. In this chapter, we'll look at some of the third-party tools available to analyze autosomal DNA (atDNA).

What are Third-Party Tools?

Each of the major testing companies—23andMe **<www.23andme.com>**, AncestryDNA **<www.dna.ancestry.com>**, and Family Tree DNA **<www.familytreedna.com>**—offer tools that the test-taker can utilize. However, several programmers and genetic genealogists have created third-party DNA tools and applications that are independent of the testing companies and offer additional capabilities and analyses. Born from a desire to extract every bit of information from DNA test results, these third-party tools offer the only way to compare raw data from one company (i.e., the test-taker's DNA sequence) to raw data from another company (provided that both people have uploaded their raw data to the same third-party tools). Two of the most commonly utilized tools are GEDmatch <**www.gedmatch.com**> and DNAGedcom <**www.dnagedcom.com**>.

This section explores the use of some of these third-party DNA tools.

RESEARCH TIP:

Mind Your Privacy

Although the third-party sites discussed in this chapter have policies that respect and protect the user's privacy, no company, service, or website (including the testing companies) can guarantee absolute privacy. Accordingly, raw data or other test results should only be uploaded to or accessed by a third-party site if the owner of the raw data or test results has provided explicit permission. Note this also includes a person who provided the DNA sample to a third-party service but is not considered the owner of the raw data or test results.

GEDmatch

The most popular third-party tool, by far, is GEDmatch. GEDmatch was created by Curtis Rogers and John Olson using donations and their own time. In October 2015, GEDmatch reported that it "has over 130,000 registered users, over 200,000 samples in its DNA database, and over 75 million individuals in its genealogical database" <**www.genomeweb.com/informatics/consumer-genomics-third-party-tool-makers-look-develop-services-while-keeping-user**>. The two-hundred-thousand-plus samples in the database are atDNA raw data results that have been uploaded by users to GEDmatch from 23andMe, AncestryDNA, and Family Tree DNA.

The first step to using GEDmatch is to create a free account. Once you have a profile, you can access the GEDmatch tool and upload new raw data results for processing and inclusion in the database. As shown in image A, the main page of GEDmatch includes several panels, each with different information. In the File Uploads panel, you'll find links with step-by-step instructions for downloading raw data from the testing companies and uploading to the tool.

Once a raw data file is successfully uploaded to GEDmatch, it will be assigned a "kit number." Each testing company has an assigned letter that is represented in a kit number, with the first letter in a kit number representing the testing company. For example, kit *M123456* is so named because its results are from 23andMe (*M*), while kit *A123456* would be made up of AncestryDNA (*A*) results and *T123456* for Family Tree DNA results (*T*).

Some tools are available immediately for newly updated results, while raw data must be processed for one or two days before it is available for other tools. Genetic genealogists interested in learning more about their atDNA test results should experiment with the tools at GEDmatch and keep checking back as the site continues to grow and develop new tools and functionality.

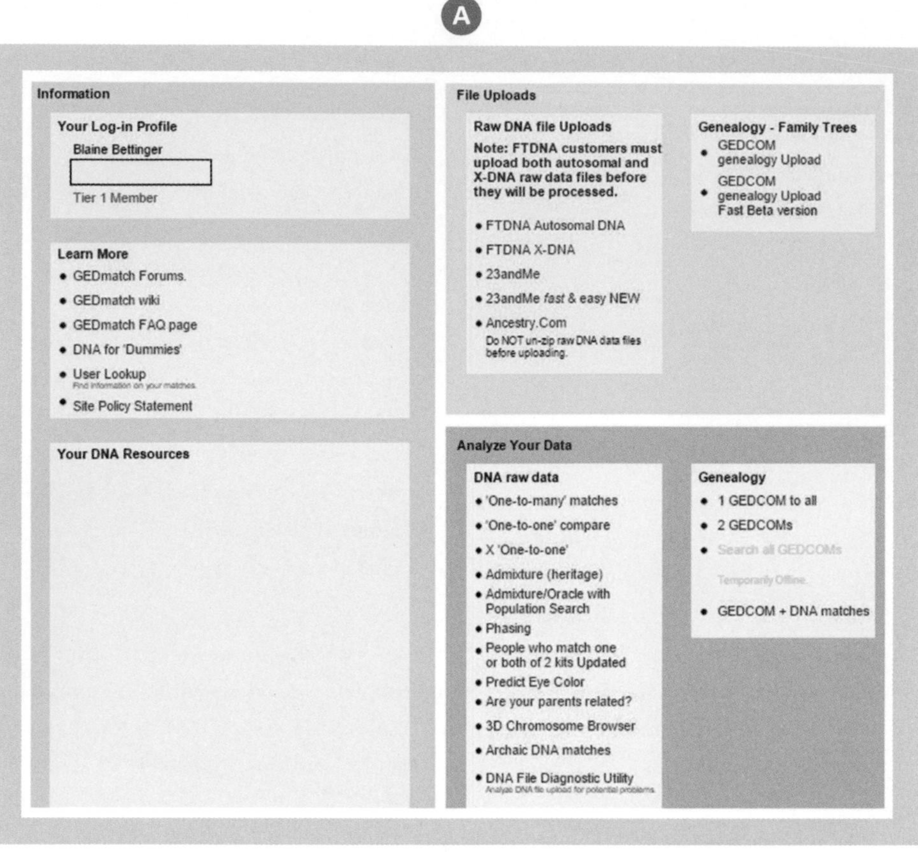

GEDmatch has a number of tools you can use to analyze your atDNA test results.

There are many free tools available at GEDmatch, some of which we'll discuss in depth later. The most important and most commonly used are:

- **"One-to-many" matches:** These compare the raw data of a single kit to the raw data of every other kit in the GEDmatch database (120,000 and growing) in order to identify genetic cousins who share an amount of DNA above the sharing threshold. The sharing threshold, which can be manually adjusted higher or lower, is 7 centimorgans (7 cMs), meaning that two individuals must share a segment of DNA that is 7 cMs or longer in order to be identified as a genetic cousin using the One-to-many tool.
- **"One-to-one" compare:** This compares the atDNA data of a single kit to the atDNA data of one other kit in order to identify segments of atDNA shared between the kits above the sharing threshold, if any. The user can manually adjust the sharing threshold to be higher or lower than the default 7 cMs.

- **X "One-to-one":** This compares the X-chromosomal DNA (X-DNA) data of a single kit to the X-DNA data of one other kit in order to identify segments of X-DNA shared between the kits above the sharing threshold, if any. The user can manually adjust the sharing threshold to be higher or lower than the default of 7 cMs.
- **Admixture:** In this process, the program performs an ethnicity analysis of atDNA data using one of several different proprietary ethnicity calculators. Results can be provided in several different formats, including as percentages, in a chromosome browser, or as a pie chart, among others.
- **People who match one or both of 2 kits:** This uses two kit numbers to identify genetic cousins above a sharing threshold in three different categories: (1) Kits in the GEDmatch database that match both of the two entered kit numbers; (2) Kits in the GEDmatch database that match only the first of the two entered kit numbers; and (3) Kits in the GEDmatch database that match only the second of the two entered kit numbers.
- **Are your parents related?:** This determines whether the atDNA data of a kit has any segments of DNA that are the same from both parents, meaning both copies of a chromosome have the same DNA—and were inherited from the same ancestor—at that location. This can occur, for example, if the parents are related.

In addition to the free tools at GEDmatch, you can purchase a group of applications called Tier 1 Tools for a ten-dollar donation for each month of use. These tools are designed for more advanced users:

- **Matching Segment Search:** This organizes and displays in a graphic all the segments of DNA that a kit shares with other kits in the GEDmatch database.
- **Relationship Tree Projection:** This calculates probable relationship paths between two GEDmatch kits based on atDNA and X-DNA sharing and genetic distances. This is a highly experimental tool and should be utilized cautiously.
- **Lazarus:** This app creates a surrogate kit that represents a recent ancestor. DNA segments for the surrogate kits are found by comparing the DNA of descendants of the recent ancestor (Group 1) to the DNA of non-descendant relatives of the ancestor (Group 2). Any segments of DNA shared between Group 1 and Group 2 are assigned to the surrogate kit of the recent ancestor.
- **Triangulation:** This tool identifies "triangulation groups" from among the matches of a GEDmatch kit above the sharing threshold, the default of which is 7 cM. A "triangulation group" is a group of three or more GEDmatch kits that all share a segment of DNA in common with each other.

B

					Haplogroup		Autosomal				X-DNA				
Kit Nbr	Type	List	Select	Sex	Mt	Y	Details	Total cM	largest cM	Gen	Details	Total cM	largest cM	Name	Email
▼ ▲					▼ ▲	▼ ▲		▼	▼	▼ ▲		▼	▼	▼ ▲	▼ ▲
	F2	L		F			A	785.8	67.9	2.1	X	72.5	25.6		
	F2	L		F	A2w	G	A	743	63.3	2.1	X	105.8	84.5		
	F2	L		F	A2w		A	710.1	52.6	2.2	X	113.3	69.3		
	F2	L		M		G	A	712.5	49.6	2.2	X	112.8	69.1		
	F2	L		M		G	A	574.8	68	2.3	X	89.5	69.1		
	V3	L		F	A2		A	595.9	48.6	2.3	X	122.6	84.5		
	F2	L		M	A2	R1b	A	300.3	47.6	2.8	X	43.5	17		
	V2	L		M	A2	R1b	A	300	47.6	2.8	X	27.6	15.5		
	F2	L		M	H	R1b	A	123	47.6	3.4	X	0	0		
	V4	L		U			A	117.8	29.1	3.5	X	0	0		
	F2	L		F			A	97.8	25.1	3.6	X	5.8	5.8		
	F2	L		F			A	97.8	25.1	3.6	X	5.8	5.8		
	F2	L		M	L2b	R-M269	A	87.7	25.1	3.7	X	0	0		
	F2	L		M	H	R1b	A	85.3	24.4	3.7	X	0	0		
	F2	L		F			A	70.2	32.2	3.8	X	17.3	17.3		
	F2	L		M			A	60.8	15.9	3.9	X	0	0		
	F2	L		F			A	55.7	25.2	4	X	0	0		
	V4	L		M	HV4	R1b1b2a1a	A	56.2	20.9	4	X	0	0		

A One-to-many analysis compares your data to that of all other GEDmatch users.
Kit numbers, names, and e-mail addresses have been removed for privacy.

The One-to-Many Matches Tool

Since the One-to-many tool compares the atDNA data of a kit to every other kit in the GEDmatch database, it lets you "go fishing" in the pools of other companies without testing there. Indeed, a third-party tool is the only way to compare the raw DNA data from one company to the raw DNA data of another company. Using this tool, you can identify up to fifteen hundred genetic cousins from the GEDmatch database.

One of the benefits of the One-to-many tool is that users can adjust the settings. Although the default threshold for identifying a genetic cousin is at least one segment of 7 cMs or greater, users can decrease this to 3 cMs or increase it to 30 cMs. Decreasing the threshold will increase the number of identified genetic cousins (up to fifteen hundred), while increasing the threshold will decrease the number of identified genetic cousins.

The One-to-many analysis creates a table of every kit in the database that shares a segment of DNA with the query kit, ranked from the kit that shares the most DNA to the kit that shares the least DNA down to the sharing threshold (image B). Each row in the table is a kit that shares DNA with the query kit. Each row provides: the sex of the kit owner; a mitochondrial DNA (mtDNA) and/or Y-chromosomal DNA (Y-DNA) haplogroup if the owner of that matching kit has provided that information; the total amount of DNA shared between the two kits; the largest segment of DNA shared between the two kits; an estimate of the number of generations between the two kits; the total amount of X-DNA shared between the two kits (if any); the largest segment of X-DNA shared between the two kits (if any); and the kit owner's e-mail address.

Since the e-mail address for each match is provided, you can contact other users to identify the shared ancestry with that match. Additionally, you can compare the total amount of DNA shared with a match to published relationship estimates in order to guess the possible relationship with that match. As discussed in chapter 6, the International Society of Genetic Genealogy's Wiki page "Autosomal DNA Statistics" **<www.isogg.org/wiki/Autosomal_DNA_statistics>** includes a table showing the predicted amount of total shared DNA for a wide variety of different relationships.

The One-to-One Compare Tool

The One-to-one tool compares the atDNA data of a single kit (the "query kit") to the atDNA data of one other kit in order to identify each segment of atDNA shared between

C

Minimum threshold size to be included in total = 700 SNPs
Mismatch-bunching Limit = 350 SNPs
Minimum segment cM to be included in total = 7.0 cM

Chr	Start Location	End Location	Centimorgans (cM)	SNPs
1	242,558,207	247,169,190	8.7	1,159
2	10,942,071	16,659,951	11.1	1,504
3	36,495	2,922,575	8.0	1,170
4	29,056,005	40,396,996	12.5	2,375
4	87,070,584	107,109,819	15.5	3,782
4	160,107,672	178,004,613	19.5	3,665
5	163,444,318	169,013,893	10.3	1,525
6	148,878	6,003,774	17.2	2,015
8	22,256,093	38,441,503	18.2	3,744
9	85,487,489	91,480,694	10.2	1,622
13	34,150,290	76,233,170	39.9	10,411
16	22,904,565	62,443,241	37.1	6,271
16	78,825,386	85,240,531	22.6	3,317
17	45,469,867	74,069,538	47.3	7,144
18	11,769,857	36,203,700	23.1	5,267
21	31,141,929	35,847,942	8.7	1,363
22	43,902,055	49,528,625	18.8	2,127

Largest segment = 47.3 cM
Total of segments > 7 cM = 328.8 cM
Estimated number of generations to MRCA = 2.7

A One-to-many analysis compares your data to that of all other GEDmatch users.

the kits above the sharing threshold, if there are any such segments. The sharing threshold can be manually adjusted by the user to be higher or lower than the default of 7 cMs.

The One-to-one tool creates either a table of shared segments or a graphic display of shared segments. Image C shows the table of segments of DNA shared by first cousins once removed, with the sharing threshold set to 7 cMs. These first cousins once removed share twenty-two segments of DNA, ranging from a maximum of 47.3 cMs to a minimum 8.0 cMs. For each shared segment, the One-to-one tool provides the chromosome where the shared segment is located, and the start and stop location of that segment on the chromosome.

The same information can be provided in a chromosome browser, which (as discussed in chapter 6) shows where along each of the chromosomes a shared segment is located. In image D, the same first cousins once removed are compared with the sharing threshold set to 7 cMs. The segments underlined with a blue bar are the segments of DNA above the matching threshold and shared by the first cousins once removed.

D

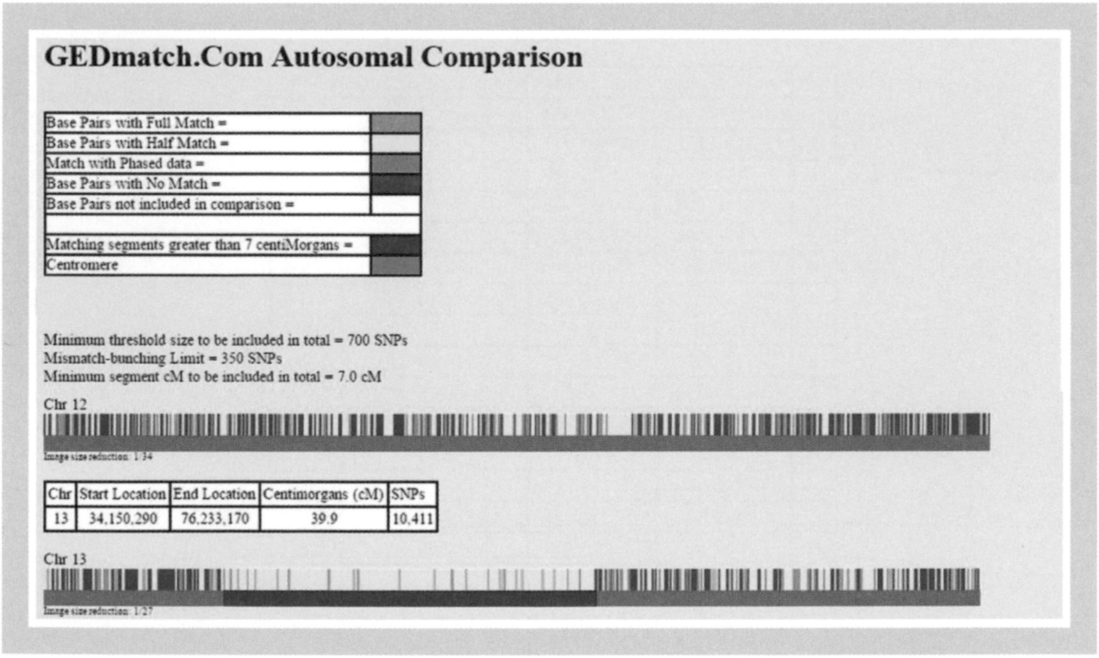

GEDmatch.Com Autosomal Comparison

Base Pairs with Full Match =	
Base Pairs with Half Match =	
Match with Phased data =	
Base Pairs with No Match =	
Base Pairs not included in comparison =	
Matching segments greater than 7 centiMorgans =	
Centromere	

Minimum threshold size to be included in total = 700 SNPs
Mismatch-bunching Limit = 350 SNPs
Minimum segment cM to be included in total = 7.0 cM

Chr 12

Image size reduction: 1/34

Chr	Start Location	End Location	Centimorgans (cM)	SNPs
13	34,150,290	76,233,170	39.9	10,411

Chr 13

Image size reduction: 1/27

You can view your One-to-one comparison in a chromosome browser. Yellow indicates portions of chromosome that the test-taker and match share on one copy of that chromosome (half match), while green indicates where the two share DNA on both chromosomes (full match). Red indicates a base pair that the test-taker and match don't share on either copy of a chromosome. (Note the report can generate all twenty-two chromosomes; this image shows only chromosomes 12 and 13 for space's sake.)

E

Chr	Start Location	End Location	Centimorgans (cM)	SNPs
21	9,849,404	24,863,804	25.1	2,989
21	34,176,163	46,909,175	31.4	4,220

Chr 21

Image size reduction: 1/10

Siblings should share large portions of both of their chromosomes. The blue bar, as well as the green and yellow, indicate where these two brothers share large portions of their chromosome 21.

Although current chromosome browsers only show one chromosome, remember that a test-taker actually has two chromosomes: one from his mother and one from his father. A "half match"—shown in yellow—indicates DNA matching on one of the chromosomes at that location. Which chromosome it is cannot be determined without more information, and in this case it is the paternal chromosome because this is a paternal first cousin once removed.

If any of the segments were a "full match"—shown in green—the first cousins once removed would share segments of DNA on *both* copies of their chromosome. This is most commonly seen in full-sibling comparisons, as shown in image E. On chromosome 21, these brothers share three segments of DNA underlined by the blue bars. Although there are only two blue bars, a portion of the blue bar on the left side of the chromosome includes a full match—again, shown in green—where both brothers inherited DNA from each parent. Comparing the table to the graphic display, however, shows that GEDmatch only provides the start and stop positions of half segments.

The X One-to-One Tool

The X One-to-one tool compares the X-DNA data of a single kit (the "query kit") to the X-DNA data of one other kit in order to identify each segment of X-DNA shared between the kits above the sharing threshold, if there are any such segments. The user can manually adjust the sharing threshold to be higher or lower than the default 7 cMs. The output of the X One-to-one tool is either a table of shared segments or a chromosome browser display of shared segments, similar to the atDNA One-to-one tool.

The Are Your Parents Related Tool

The Are your parents related? tool determines whether the atDNA data of a kit has any segments of DNA that are the same from both parents, meaning both copies of a chromosome have the same DNA (i.e., inherited from the same ancestor) at that location.

F

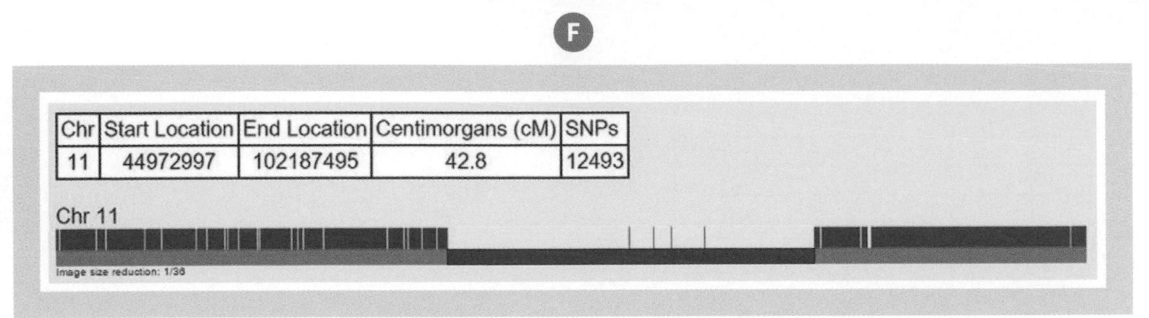
Chr	Start Location	End Location	Centimorgans (cM)	SNPs
11	44972997	102187495	42.8	12493

GEDmatch has a tool that will help you determine if your parents are related. Results like these suggest that the test-taker's parents both inherited the DNA indicated in yellow (called a Run of Homozygosity, or ROH) from a common ancestor.

Segments of shared DNA on both chromosomes are called Runs of Homozygosity (ROH). This can occur, for example, if the parents are related. The results of the analysis are presented in a chromosome browser (image F), with any ROH above 7 cMs shown in yellow and underlined by blue.

It is not uncommon for individuals to share one or two small segments of DNA from both parents, which means that the parents were likely distantly related. In some populations, however, where there has been marriage and reproduction by relatives, it is more common to have these ROH.

DNAGedcom

DNAGedcom is another third-party tool commonly used by genetic genealogists (image G). The site was founded by Rob Warthen and launched in February 2013, and its tool allows for download of important data files from 23andMe and Family Tree DNA. It also has third-party tools for GEDCOM comparisons, in-common-with analysis, and triangulation. According to the creators of DNAGedcom, the goal of the site "is to reduce the human involvement in extracting, measuring the data, to provide software for solutions for DNA matching from results and to determine relationships from this data and family trees and to provide additional metrics and comparisons not now available to the user" <www.dnagedcom.com/FAQ.aspx>.

The programmer behind DNAGedcom is constantly improving existing tools and developing new ones. As with GEDmatch, it is important that genetic genealogists monitor this and other third-party tools to stay abreast of developments and new tools.

The first step to using DNAGedcom is to create a free account. Once a user has a profile, he can access the DNAGedcom tools, including each of the following (many of which are discussed in more detail in chapter 10):

- **23andMe data download:** You can download 23andMe data, including a spreadsheet of matches, via the DNAGedcom Client. The DNAGedcom Client is an appli-

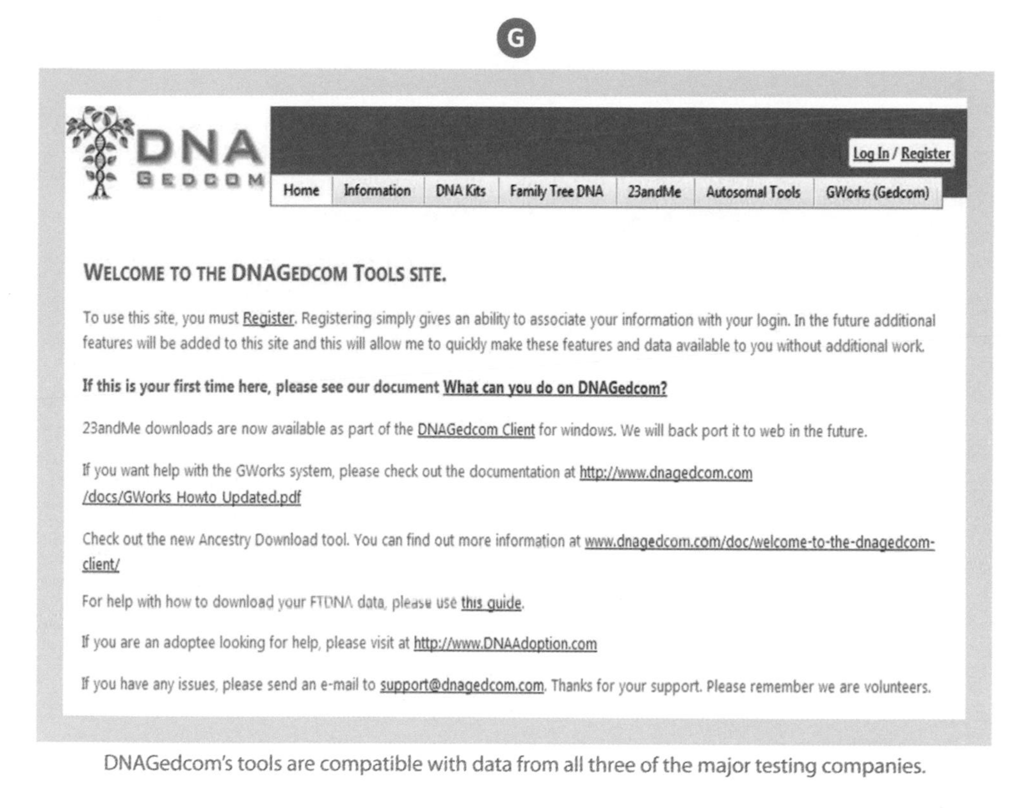

DNAGedcom's tools are compatible with data from all three of the major testing companies.

cation that runs on your machine and is only available to subscribers (Members > Subscriber Information).

- **AncestryDNA data download:** You can download AncestryDNA data into spreadsheets using the DNAGedcom Client, including match lists (with total shared cM values), the ancestors of your matches, and ICW lists.
- **Family Tree DNA data download:** You can download atDNA files from Family Tree DNA, including the Family Finder match list, all chromosome browser data, and the In Common With (ICW) information. A copy of the data is automatically saved in your Member folder (Members > View Files) at DNAGedcom. The downloaded match data from Family Tree DNA can also be automatically uploaded to the GWorks tool (more about that in a bit).
- **Autosomal DNA Segment Analyzer (ADSA):** This utilizes Family Tree DNA or GEDmatch data to generate tables in your browser that include match information, segment information, and ICW information. The tool is then used to triangulate matching segments among groups of three or more people, although it does not provide perfect triangulation since it relies only on ICW information.

- **Gedmatch Data Uploader:** This accepts the results of the Matching Segment Search and Triangulation Tier 1 tools at GEDmatch. See <**www.dnagedcom.com/docs/GEDmatchADSA.pdf**> for more information. The uploaded results can then be used for DNAGedcom's ADSA, JWorks, and KWorks tools.
- **JWorks:** This downloadable Excel tool generates a spreadsheet of overlapping segments and ICW status among matches, which helps identify potential triangulation groups. The tool requires three things: (1) chromosome browser data (segment data); (2) full match list; and (3) ICW status.
- **KWorks:** This generates a spreadsheet of overlapping segments and ICW status among matches, which helps identify potential triangulation groups. KWorks is the online version of JWorks, and just like JWorks, the tool requires three components: (1) chromosome browser data (segment data); (2) full match list; and (3) ICW status.
- **GWorks:** This compares family tree information to identify shared ancestors. GWorks can also sort and filter tree information and perform Boolean searches of the trees. The tool can use GEDCOMs uploaded by the user, family tree information downloaded from matches at AncestryDNA using the DNAGedcom Client (or the AncestryDNA Helper tool, another third-party tool available to test-takers), and family tree information downloaded from matches at Family Tree DNA using DNAGedcom's Download Family Tree DNA Data tool (Family Tree DNA > Download Family Tree DNA Data). For more information about GWorks, see <**www.dnagedcom.com/docs/GWorks_Howto_Updated.pdf**>.

Autosomal DNA Segment Analyzer (ADSA)

The Autosomal DNA Segment Analyzer (ADSA) is a tool that takes data from Family Tree DNA or GEDmatch and generates an online table that includes the test-taker's match information, segment information, and color-coded ICW information that facilitates triangulation (image H). The ADSA manual can be found at <**www.dnagedcom.com/adsa/adsamanual.html.php**>.

Each match is mapped to the chromosomes, with overlapping segments placed adjacent to each other (image I). If you hover over the shared segments table, the tool provides information such as surnames, suggested relationships, and matching segments. You can run the tool for a single chromosome or all chromosomes, and the minimum matching segment size can be raised or lowered (although a minimum of 7 cMs is strongly suggested by DNAGedcom to keep the output manageable and reliable).

Note that this is *pseudo*-triangulation, not actual triangulation. True triangulation requires information about whether an apparently overlapping segment is actually shared

H

You may select a report type in the drop down menu below which will change the input options.

Classic ADSA

	Chromosome to Graph (1-22, X, or blank=all) Base Pairs: [ ] to [ ]
7	Minimum Segment Length in cM (5 or greater STRONGLY recommended)
500	Minimum SNPs in a segment Display raw data in a table so it can be copied to a spreadsheet
500	Width of segment graph in pixels (Do not check if you want full formatting)

[REQUIRED] Kit number of the data to be used to construct the report.

The Autosomal DNA Segment Analyzer (ADSA) will triangulate matching segments among three or more test-takers.

I

CHROMOSOME 13
18 matching segments
Longest is 126.40 cM, Graph = 192 KBP/pixel

MATCH NAME	START	END	cM	SNPS	EMAIL	ICW	SEGMENTS
Ki	17956717	26574412	17.37	2904			17.37
St	17956717	27303766	18.78	3104			18.78
Br	17956717	29976511	23.50	4004			23.50
El	17956717	114121631	126.48	27967			126.48
Lc	17956717	114121631	126.48	27967			126.48
M	17956717	114121631	126.48	27967			126.48
M	17956717	114121631	126.48	27967			126.48
D	31046627	74674455	42.43	11400			42.43
St	31046627	114121631	101.14	23563			101.14
Ki	32469078	75774139	42.50	11400			42.50
D	33949169	76154533	40.45	11000			40.45
M	43809556	71267457	21.16	6400			21.16
M	46704832	77940169	25.63	7700			25.63
Ty	49744749	77940169	23.79	7000			23.79
Br	89819196	113503268	46.32	8526			40.32
M	94507381	104589639	19.76	3800			19.76
St	97791131	114121631	37.79	6063			37.79
Al	107545226	114121631	16.39	2263			16.39

ADSA results will tell you how much DNA you share with other users, but it can't pinpoint exactly which DNA you share. Names and e-mail addresses of matches have been removed for privacy.

in common, not just that two people share DNA in common. In ADSA and similar tools, the test-taker only knows from ICW information that person A, person B, and himself all share some DNA in common; it isn't known exactly *which* segment(s) person A and person B share. Accordingly, person A and person B might share the identified segment in common (which, from my experience, is common), or they might share a completely different segment of DNA in common. Regardless, the ADSA tool is very useful for identifying potential triangulation groups that can then be explored by contacting the members of the group.

	Chr	Start	Stop	cM	Abe	Ben	Cara	Donna	Eden	Frank	Gill	Hera	Ira	Jack	Kim	Lea	Mia
Abe	13	17956717	114121631	126.48		X	X	X	X	X	X	X					
Ben	13	17956717	114121631	126.48	X			X				X	X	X		X	X
Cara	13	17956717	114121631	126.48	X			X	X	X	X						
Donna	13	17956717	114121631	126.48	X	X	X		X	X	X	X		X	X	X	X
Eden	13	31046627	74674455	42.43	X		X	X		X	X						
Frank	13	31046627	114121631	101.14	X		X	X	X		X						
Gill	13	32469078	75774139	42.5	X		X	X	X	X							
Hera	13	33949169	76154533	40.45	X	X		X						X		X	X
Ira	13	39986582	47064783	8.32		X											
Jack	13	43809556	71267457	21.16		X		X				X			X	X	X
Kim	13	44812957	60049758	10.08				X						X		X	
Lea	13	45197769	58723000	9.07		X		X				X		X	X		
Mia	13	46704832	77940169	25.63		X		X				X		X			

KWorks can export matches as an Excel spreadsheet. An *X* indicates when two individuals share ancestors in common on chromosome 13.

KWorks

The information generated by the KWorks tool is the same as the information generated by the ADSA tool, although KWorks produces a different visual display. In contrast to the color-coded output of ADSA, KWorks creates a spreadsheet of potential triangulation groups using ICW data, segment data, and match lists. The tool requires an ICW file and a segment file, and generates the downloadable spreadsheet.

In image J, the *X* indicates ICW status and thus these individuals on chromosome 13 are grouped into potential triangulation groups. For more about the JWorks and KWorks tools, see <**www.dnagedcom.com/JWorks/Jworks_Kworks.pdf**>.

Other Tools

In addition to GEDmatch and DNAGedcom, there are many other third-party tools that genealogists can use to maximize the genetic genealogy experience. Here's a list of some of the most common third-party tools for atDNA:

- **David Pike's Utilities** <**www.math.mun.ca/~dapike/FF23utils**> is a free comprehensive suite of tools for several advanced phasing and analyzing raw data, including searching for ROHs and searching for shared DNA in two files. Unlike other third-party tools, David Pike's Utilities operates within your browser, which may alleviate some privacy concerns of people hesitant to upload raw data to a third-party site.
- **DNA Land** <**dna.land**> is a free tool for analyzing ethnicity and finding genetic cousins. The tool is run by academics from Columbia University and the New York Genome Center.

- **Genetic Genealogy Tools** <www.y-str.org> comprises an impressive and ever-growing list of advanced tools for analyzing raw data, including an X-DNA Relationship Path Finder, Ancestral Cousin Marriages, Autosomal Segment Analyzer, a DNA Cleaner, an SNP Extractor, My-Health, and many more.
- **Genome Mate Pro** <www.genomemate.org> is an extremely powerful, free computer program that organizes data from 23andMe, AncestryDNA, Family Tree DNA, and GEDmatch, among other sources, into a single working file. Information is stored locally on your computer, which helps maintain the privacy of your data.
- **Promethease** <www.promethease.com> is a literature retrieval system that creates a personal DNA report based on scientific literature and the test-taker's raw data files from 23andMe, AncestryDNA, and Family Tree DNA. Reports contain information about health and ancestry as well as several other new options. Promethease has a variable cost depending on which raw data files are used, and how many different raw data files are analyzed at once.
- **Segment Mapper** <www.kittymunson.com/dna/SegmentMapper.php> is a free, powerful mapping tool that shows specific DNA segments in a graphic chromosome-style chart.

Also see the impressive list of free and paid third-party tools available on the International Society of Genetic Genealogy's Wiki <www.isogg.org/wiki/Autosomal_DNA_tools>.

CORE CONCEPTS: THIRD-PARTY AUTOSOMAL-DNA TOOLS

- Many different free and paid third-party tools are available to atDNA test-takers.
- GEDmatch <www.gedmatch.com> is the most popular third-party site and offers many different tools for users, including the ability to find genetic cousins who may have been tested at a different testing company.
- DNAGedcom <www.dnagedcom.com> is a popular third-party site that provides powerful data collection and analysis tools to test-takers.
- Before using a third-party tool, consider potential privacy issues that might be raised. Additionally, have the person who provided DNA grant permission before his raw data is uploaded to a third-party site.

Getting Started with Third-Party Programs Checklist

With so many third-party tools, it can be difficult to know which to use and how they might be useful. If you are interested in experimenting with these tools—and you are comfortable analyzing your raw data (including possibly uploading your raw data to the website)—then here are some steps every new test-taker should take:

☐ **Download your raw data from the testing company.** Choose just one testing company if you've tested at two or three. As discussed previously in the chapter, you can find links with step-by-step instructions for downloading raw data from each of the testing companies at GEDmatch in the panel labeled *File Uploads.* AncestryDNA or Family Tree DNA raw data may be preferable if you'd like to avoid potentially sharing health information.

☐ **Create a free profile at GEDmatch.** Upload the raw data. Now you can use any of the free tools available at GEDmatch.

☐ **Run the DNA File Diagnostic Utility.** Use this to ensure your kit was properly uploaded and processed. Since it takes some time (usually hours or a day or two) to completely process a kit, you may have to wait to perform this analysis. Look for any red warning signs that your kit was not properly processed. Follow the directions provided, or delete your kit and reload the raw data.

☐ **Run the Are Your Parents Related? tool.** I recommend that this test be utilized for every kit uploaded to GEDmatch, as this will reveal whether there is significant DNA shared on both sides of the family. Finding that a test-taker's mother and father share DNA means that they share ancestry, and could have a strong impact on subsequent genealogical studies. Most kits, however, will report that there are "no shared DNA segments found."

☐ **Run the One-to-many matches tool.** Do this to search for genetic relatives at GEDmatch, particularly if you haven't tested at all three companies. I recommend that for your initial search raise the threshold to 15 cMs or higher (the default is 7 cMs) since you're going to be focusing only on the very closest matches first.

Once you've mastered these steps, you're ready to explore the other tools at GEDmatch, as well as other third-party tools.

9

Ethnicity Estimates

How reliable is a prediction of 37-percent British ancestry? Why doesn't your confirmed German or Italian ancestry show up in your ethnicity prediction? You can answer these questions by learning more about the ethnicity estimates that each of the major testing companies provides with the test-takers' autosomal DNA (atDNA) testing results. While some people assume these estimates are foolproof, ethnicity estimates are merely percentages of the test-taker's DNA determined by the testing company's algorithm to be associated with a particular continent, region, or country. Unfortunately, ethnicity prediction is still a young and developing science, and these ethnicity estimates are subject to limitations that minimize their applicability to genealogical research.

What are Ethnicity Estimates?

Ethnicity estimation—also known as *admixture* or *biogeographical* estimation—is the process of assigning a test-taker's DNA to one or more populations around the world based on computerized comparisons of those segments to reference populations. Individual segments of the test-taker's DNA are assigned to the reference population that

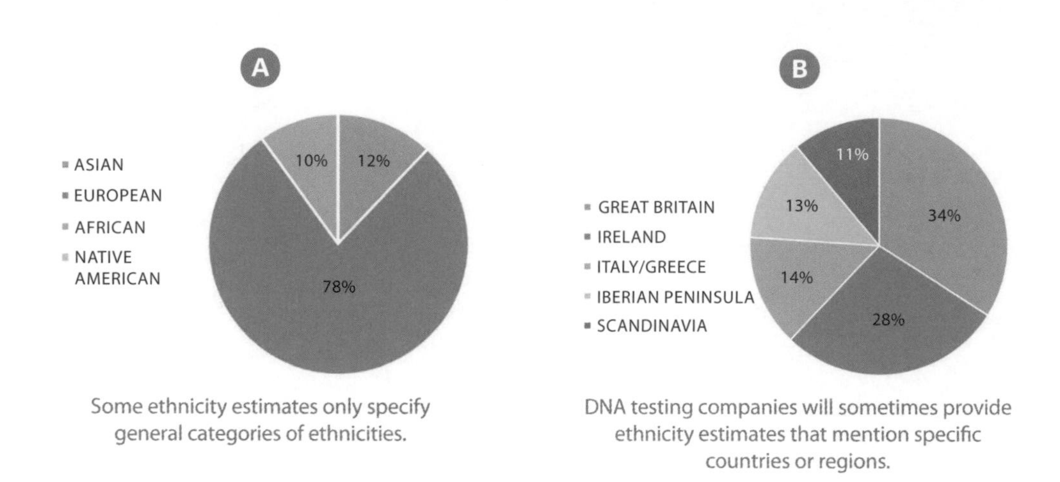

Some ethnicity estimates only specify general categories of ethnicities.

DNA testing companies will sometimes provide ethnicity estimates that mention specific countries or regions.

they most closely match, based on the assumption that the DNA most likely came from that population at some recent point in time. All assignments over the test-taker's entire genome are added together to create the overall ethnicity estimate.

Let's look at a couple examples of how ethnicity estimates are presented. In image A, test-taker Jacob Armstrong tested at one of the atDNA testing companies and received an ethnicity estimate along with his list of genetic matches. The estimate provides his ethnicity for each of four broad categories: African (10 percent), Asian (12 percent), European (78 percent), and Native American (0 percent). In image B, test-taker Millie Fuller's estimate is much more specific, with 34-percent Great Britain, 28-percent Ireland, 14-percent Italy/Greece, 13-percent Iberian Peninsula, and 11-percent Scandinavia.

Typically, estimates using broader categories are more accurate, as it's easier for geneticists to distinguish between continents (e.g., European versus Asian) than it is to distinguish between modern countries (e.g., German versus French). As a result, Jacob's results—while less specific—are more likely to be correct than Millie, as Jacob's results only suggest this DNA matches a particular *continent*, rather than a particular country or region.

Reference Populations or Panels

As discussed in earlier chapters, DNA analysis often relies on comparisons between a test-taker's DNA and a reference sample. Ethnicity estimates operate in a similar way, as geneticists compare test-takers' DNA to collections of reference samples that were obtained from known locations.

The goal of most ethnicity estimates is to identify where the test-taker's DNA was found approximately five hundred to one thousand years ago. Accordingly, a perfect reference population or panel would consist of samples of DNA obtained from populations five hundred years ago. Since this is impossible, researchers normally utilize

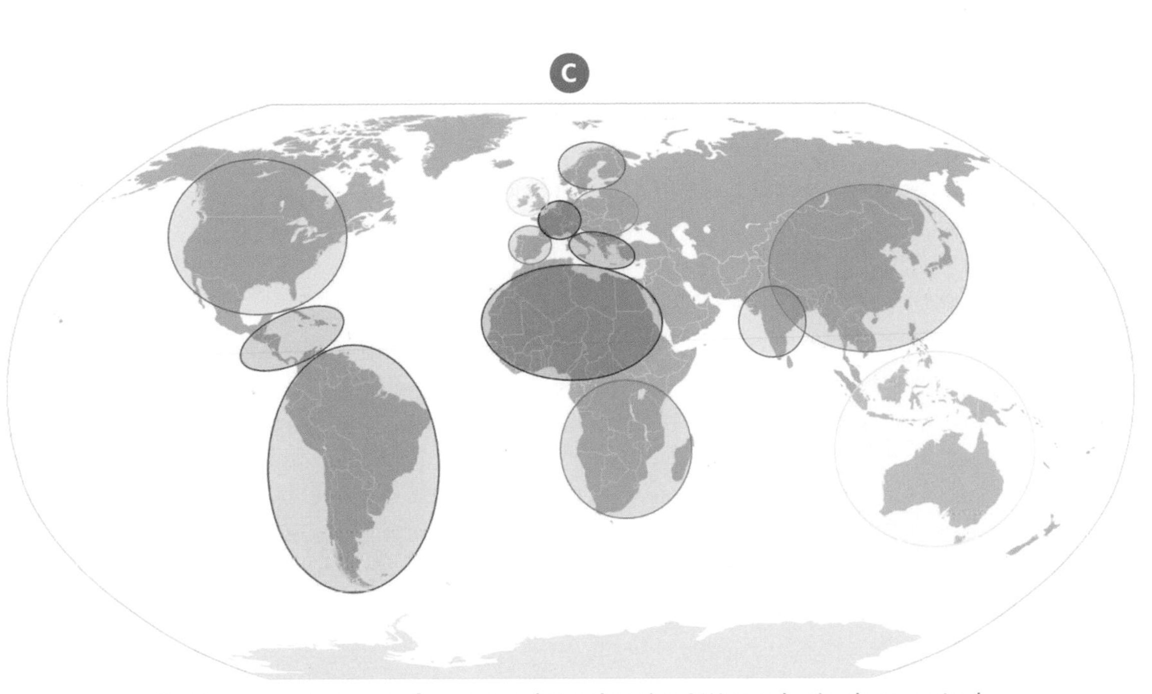

Testing companies assign reference populations based on DNA samples they have received. Because companies have sampled the DNA of more people in Europe and North America than they have in other regions, they can create more reference populations in these regions than they can in South America, Africa, Asia, or Australia and Oceania.

DNA from people who can reasonably assert that all four of their grandparents were from a specific, concentrated location (such as a county or village). While not a perfect filter, this does help formulate a more accurate reference panel.

Image C is a theoretical map showing fourteen reference populations from all over the world. Some reference populations in this map, such as those in Europe, represent relatively small regions that DNA companies feel they have adequately sampled local populations. Other regions, such as Asia, are not adequately sampled and thus the reference population represents a very large region.

Note that the size and diversity of the reference panel is a strong factor in the accuracy of an ethnicity estimate. Comparing the test-taker's DNA to a database containing only European reference populations will not, for example, produce useful results for someone with Native American or African ancestry.

Each of the testing companies has its own reference panel:

- 23andMe utilizes a database of more than ten thousand people from various populations around the world for its ethnicity reference panel, all of whom have relatively well-known ancestry. 23andMe's reference population was obtained from both 23andMe customers and from public sources <**www.23andme.com/en-int/ancestry_composition_guide**>.

- AncestryDNA uses a reference panel with more than three thousand DNA samples from people in twenty-six global regions <**www.dna.ancestry.com/resource/whitepaper/ancestrydna-ethnicity-white-paper**>.
- Family Tree DNA's reference panel is composed of numerous individuals from twenty-two different population clusters <**www.familytreedna.com/learn/ftdna/myorigins-population-clusters**>.

The companies continue to add DNA from new individuals and new populations to the reference panels.

Over the next few years, reference panels will likely continue to improve in at least two ways. First, developing reference panels will likely be populated with more individuals from a wider variety of populations. Second, reference panels will likely be populated with more ancient DNA samples being obtained from ancient remains all over the globe. Together with other improvements, these additions will help DNA testing companies significantly improve the accuracy of their ethnicity estimates.

Ethnicity Estimates from the Big Three

Together with cousin matching, ethnicity estimates are one of the two major interpretations of your DNA offered by the big three testing companies. Each of the testing companies provides an estimate of very broad regions, including Africa, Asia, the Americas, and Europe, and each attempt to break these regions down into smaller categories, often based on modern-day countries. See the Global Regions Comparison Worksheet at the end of the chapter for more detailed information.

If you test at all three companies, you should expect your ethnicity estimate to vary. In the following table are actual ethnicity estimates for the same person from each company (rounded to the nearest percentage):

Region	23andMe	AncestryDNA	Family Tree DNA
African	1%	2%	0%
Asian	0%	2%	7%
Native American	3%	3%	2%
European	96%	93%	90%

As we learned earlier in this chapter, these differences don't mean that one estimate is correct while another estimate is incorrect. Differences in the reference populations and ethnicity analysis algorithms utilized by the company will necessarily result in differences in the estimates.

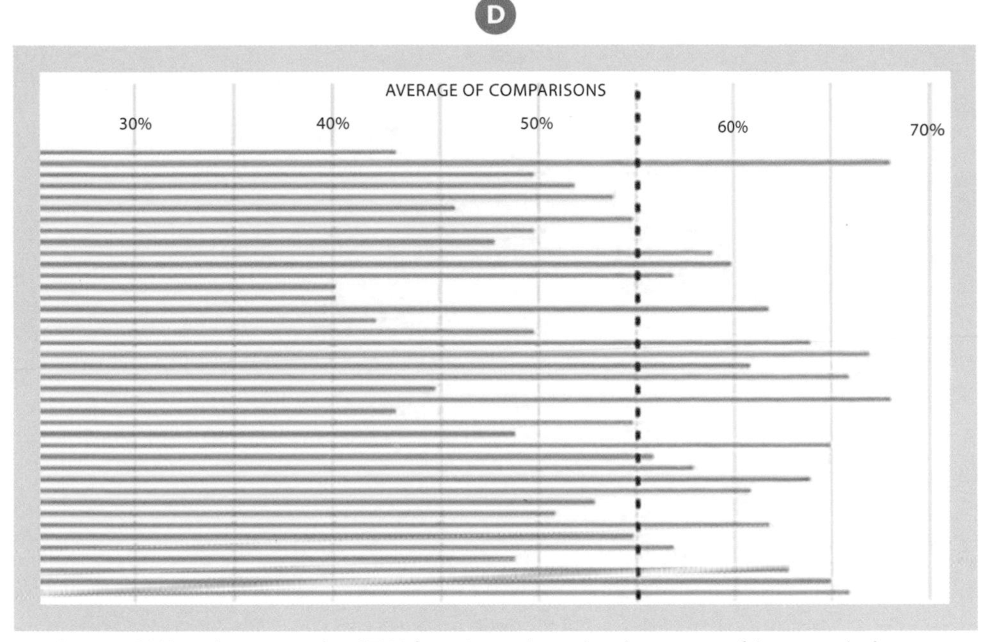

AncestryDNA analyzes test-takers' DNA forty times, then takes the average of these to calculate an ethnicity estimate for a particular reference population. For this reference population (Great Britain), the average of my comparisons was 55 percent (indicated by the dotted line).

When testing at multiple companies, remember that these differences are expected. Rather than looking for identical predictions, look for trends or patterns. According to the results in the table, for example, this person clearly has mostly European ethnicity and very likely has a significant 2- to 3-percent Native American contribution as well. The African and Asian estimates are a little more questionable, and additional research or analysis might be necessary.

AncestryDNA

Through its DNA testing, AncestryDNA provides an estimate for at least twenty-six different global regions, a number that has grown several times.

Here's how the test works. After AncestryDNA obtains the test-taker's DNA, the ethnicity algorithm performs forty different analyses using the test-taker's DNA, chopping the test-taker's DNA into random pieces for each. Running the analysis forty times allows for the program to process different combinations of the DNA and report results in a more accurate estimate while also providing a range for each estimate. Next, each of the forty analyses is compared to the reference panel to create an ethnicity estimate, and the average of the forty estimates for each region or ethnicity is determined.

In image D, for example, the average of the forty estimates for this particular ethnicity is 55 percent. Some of the forty analyses are as low as 40 percent, while others are as high as 67 percent; most of the estimates, however, fall within 45 to 65 percent. This 45 to 65 percent is the range of the estimates, which is another piece of information provided to test-takers at AncestryDNA.

AncestryDNA provides information to the test-taker in the ethnicity estimate interface. For example, my ethnicity estimate from AncestryDNA is shown in image E. The values for each region are the average obtained from the forty different analyses. Clicking on each individual region will expand that region and the range from the forty analyses is shown to the test-taker.

For much more information about AncestryDNA's ethnicity estimate, see the AncestryDNA Ethnicity Estimate White Paper **<www.dna.ancestry.com/resource/whitePaper/AncestryDNA-Ethnicity-White-Paper>**.

23andMe

Like AncestryDNA, 23andMe also starts with the test-taker's DNA sequence, then uses a proprietary computer algorithm called "Finch" to phase the DNA. **Phasing** refers to separating the test-taker's DNA sequence into the DNA provided by the mother and the DNA provided by the father. Normally, phasing is done by comparing a child's DNA to one or both of the parents' DNA. For automated phasing, however, the algorithm uses a statistical analysis to separate each parent's contribution to the test-taker's DNA. The program attempts to separate the DNA into two different contributors, but it doesn't know which contributor was the mother and which contributor was the father.

Next, 23andMe breaks the chromosomes into short, non-overlapping, adjacent segments of about one hundred markers (approximately fifty to four hundred segments for each chromosome). Each of the segments is then compared to 23andMe's reference populations to determine which of the reference populations is most similar to the segment.

The 23andMe process then corrects several different types of errors in assignments. For example, the algorithm "smooths" the data by correcting assignments that are almost certainly incorrect. If the data shows a series of ten segments in a row assigned to Population A interrupted in the middle by an assignment of a single segment to Population B, the smoothing algorithm will change the assignment to Population A. The smoothing algorithm will also correct phasing mistakes known as a "switch error" in which the phasing algorithm mixes up the DNA of one parent with the other parent. The smoothing algorithm fixes the switch error by switching the ancestry assignments back between the two versions ("mom" and "dad") of a given chromosome.

Next, 23andMe applies a confidence threshold to the data to determine what ethnicity estimates are provided to the test-taker. The test-taker is able to adjust this threshold to see estimates that are more conservative (where the threshold is higher) and estimates

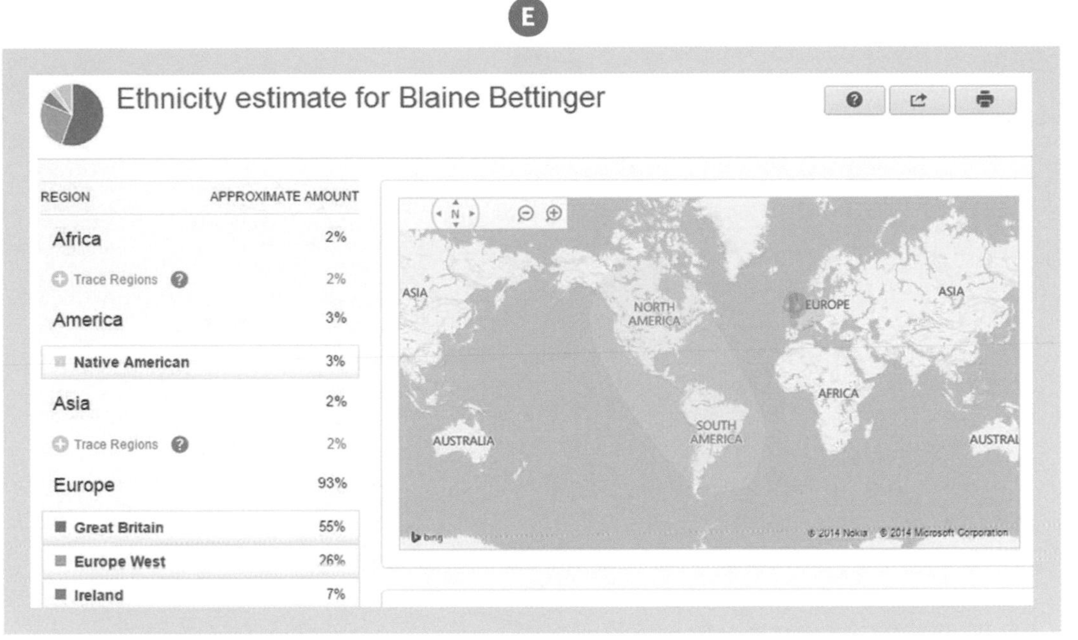

AncestryDNA compiles the averages of your estimates from each reference population and reports them alongside a map of the world showing roughly where a reference population originates from.

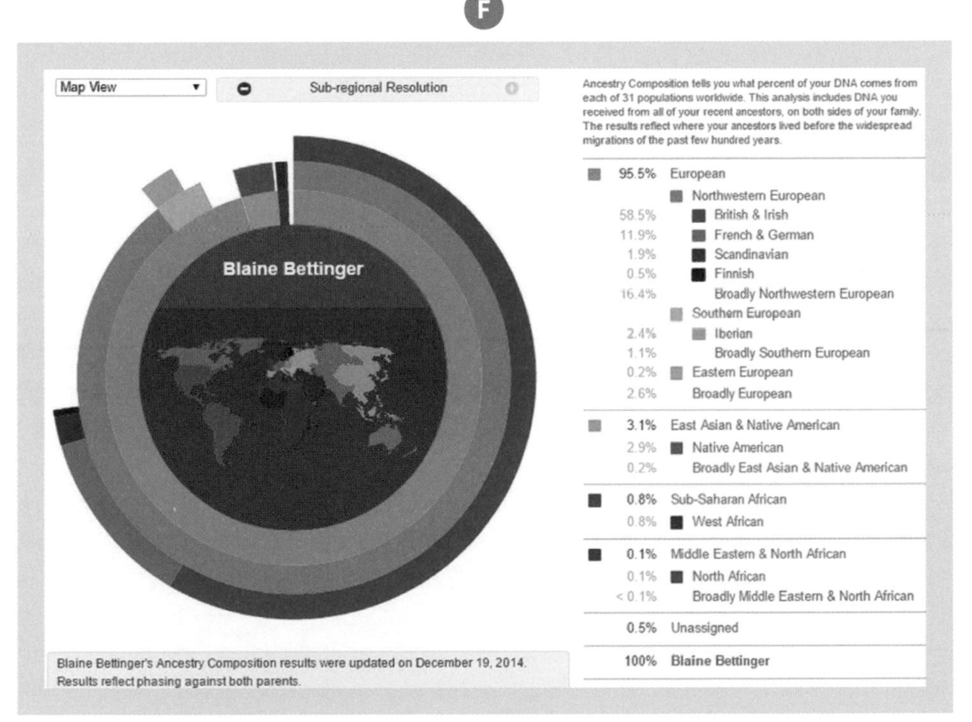

23andMe allows you to specify how you'd like to view your ethnicity estimate at different region levels, with varying degrees of confidence at each interval.

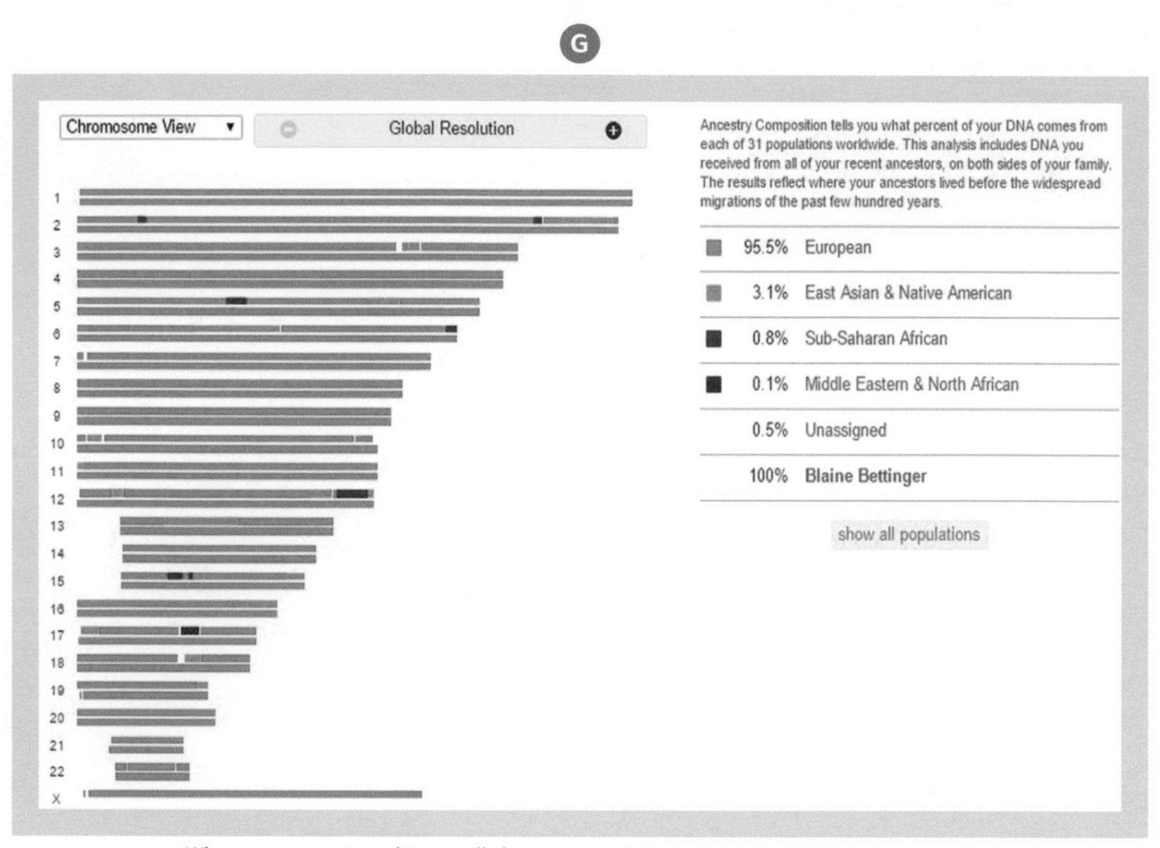

When you test at 23andMe, you'll also receive a chromosome browser that allows you to see from which region you likely received different portions of each chromosome.

that are speculative (where the threshold is lower). In 2016, 23andMe switched all users to a new user interface experience. In the old user interface, test-takers could adjust their ethnicity estimate threshold from Speculative to Standard to Conservative. In the new user interface, test-takers can adjust their ethnicity estimate threshold based on percentages ranging from 50 percent (Speculative) to 90 percent (Conservative). The default is 50 percent, and as the test-taker increases the threshold, the estimate may change as some assignments no longer satisfy the selected threshold.

My ethnicity estimate from the old user interface at 23andMe is shown in image F. The threshold for this display was set to Speculative, and the display is also adjusted to show "Sub-Regional Resolution," meaning that the ethnicity estimate identifies specific regions. For example, instead of broad categories like European or northwestern European, estimates for individual countries like British & Irish and French & German are shown.

23andMe also allows test-takers to use a chromosome browser to see where in their genome the assigned segments of each population are found (image G). The blue

(European), orange (East Asian & Native American), red (Sub-Saharan African), and purple (Middle Eastern & North African) colors on the chromosomes represent where each ethnicity assignment is found. For example, my chromosome 6 has a long orange (East Asian & Native American) segment, suggesting I received that DNA from East Asian or Native American ancestors.

Each chromosome is shown with two copies in the 23andMe chromosome browser, although there is no order to the arrangement. It is also unclear from one person's test results whether multiple segments on a chromosome all came from one parent or a mixture of the two parents. For example, chromosome 2 in the image has two small red segments and one orange segment. It is possible that the red segments came from one parent and the orange segment came from the other parent, or that one red segment came from one parent and the other red segment and the orange segment from the other parent, for example. Only if an ethnicity is identified at the same location on both chromosomes can the test taker be reasonably assured that an ethnicity came from both the mother and father. In the image, for example, most of the chromosomes are blue (i.e., European) on both copies.

To learn more about 23andMe's ethnicity estimate, see the 23andMe Ancestry Composition guide: **<www.23andme.com/en-int/ancestry_composition_guide>**.

H

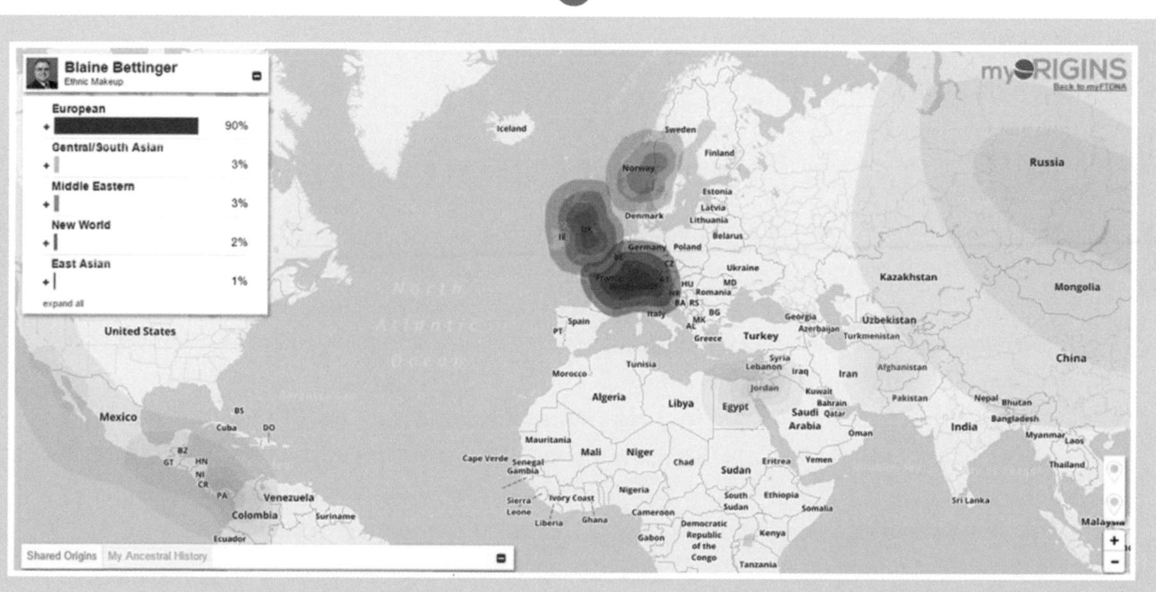

The default view in Family Tree DNA's ethnicity estimate results describes broad categories, projected over the rough regions they represent.

Family Tree DNA

Family Tree DNA's ethnicity estimate is called myOrigins, which provides an estimate for several different global regions. To do this, Family Tree DNA first obtains the test-taker's DNA sequence, then compares the DNA to the different global regions to obtain an overall ethnicity estimate.

My myOrigins ethnicity estimate from Family Tree DNA is shown in image H. The default for the user interface is to show broad categories such as European, Central/South Asian, Middle Eastern, New World, and East Asian. Clicking on these regions reveals sub-regions, as shown in image I, where expanding European to sub-regions reveals Western and Central Europe, British Isles, and Scandinavia.

I

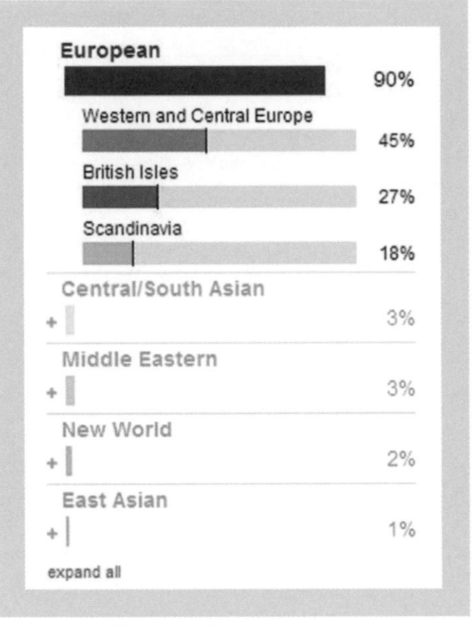

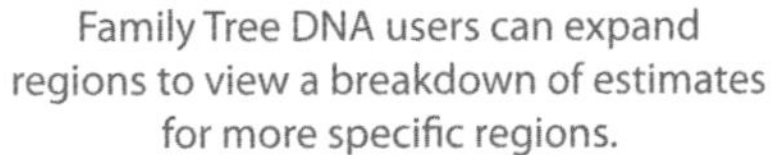
Family Tree DNA users can expand regions to view a breakdown of estimates for more specific regions.

To learn more about myOrigins, including a detailed description of the different reference populations, see the myOrigins Methodology Whitepaper **<www.familytreedna.com/learn/user-guide/family-finder-myftdna/myorigins-methodology/>**.

GEDmatch Ethnicity Calculators

In addition to the ethnicity estimates provided by the testing companies, you can also access the free third-party tool GEDmatch **<www.gedmatch.com>**, which offers test-takers a variety of different ethnicity calculators. These calculators, all created by academics and independent researchers, can help verify and expand upon your ethnicity estimates from the major testing companies.

Similar to the company ethnicity algorithms, the calculators at GEDmatch each have different reference populations. Since the different calculators at GEDmatch have different underlying algorithms and each use different reference populations, it is not unusual for ethnicity estimates to vary significantly from one calculator to another calculator, or for GEDmatch calculator estimates to vary from the ethnicity estimate from the testing companies.

Each of the ethnicity calculators at GEDmatch has two or more models the user can select from. These models are slight variations of the individual calculators and usually differ based on the composition and/or number of the reference populations used for analysis.

GEDmatch currently offers the following ethnicity calculators:

1. **MDLP** (Magnus Ducatus Lituaniae Project) **<magnusducatus.blogspot.com>** is described by the creator as a biogeographical analysis project for the territories of the former Grand Duchy of Lithuania. There are twelve different models of the MDLP Project calculators, and *World22* is the default model.
2. **Eurogenes** (Eurogenes Genetic Ancestry Project) **<bga101.blogspot.com>** focuses on European ancestry. There are thirteen models of the Eurogenes calculators, usually varying by the reference populations included in the analysis. *Eurogenes K13* is the default model.
3. **Dodecad** (Dodecad Ancestry Project) **<dodecad.blogspot.com>** focuses on Eurasian individuals. The project is named after the Greek word for "group of twelve." There are five models of the Dodecad calculator, and the default is *Dodecad V3*.

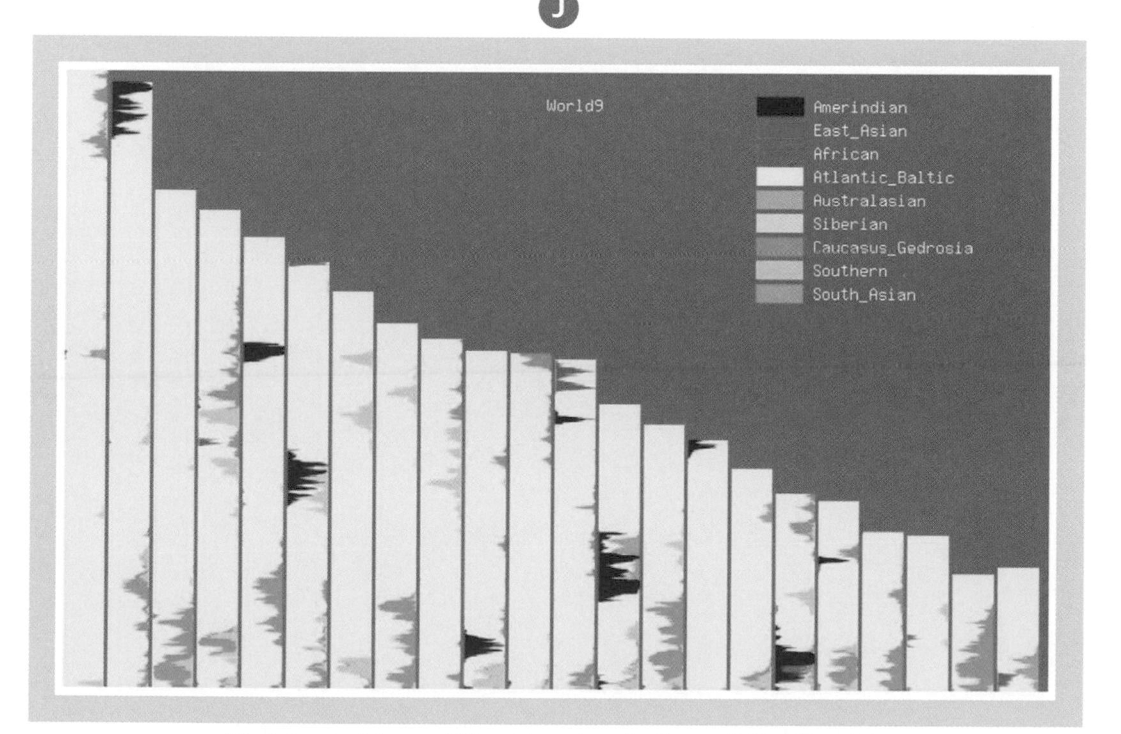

Ethnicity calculators on GEDmatch, such as the Docecad's World9 model, compare use your testing results to display your DNA in more detail.

K

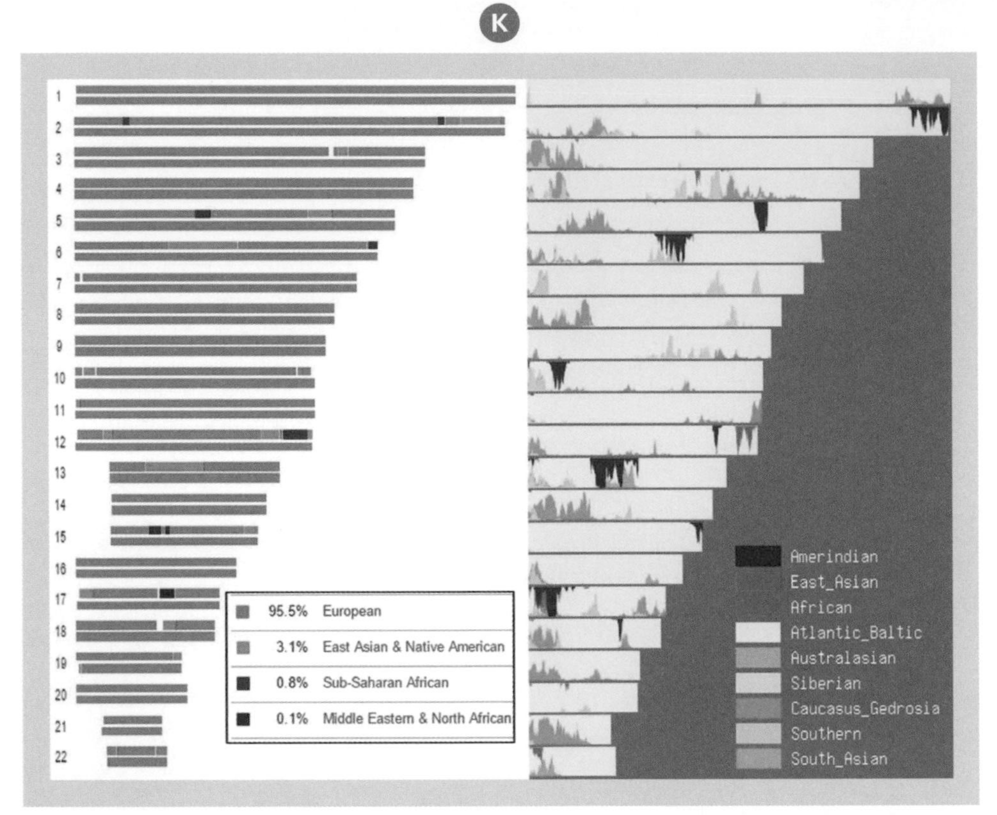

Ethnicity calculators, such as the one on the right, can correspond with (and thus verify) the estimates from your test results, such as those on the left.

L

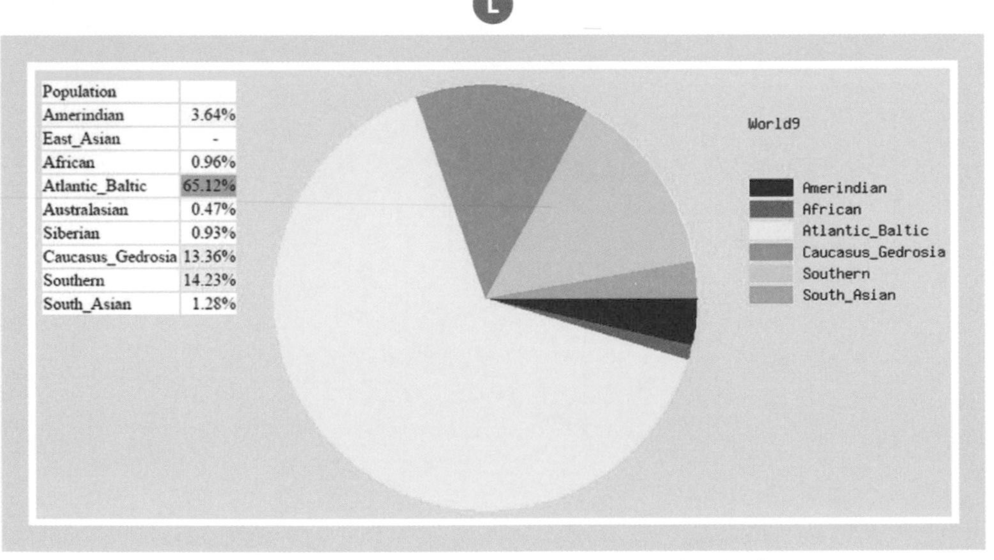

Population	
Amerindian	3.64%
East_Asian	-
African	0.96%
Atlantic_Baltic	65.12%
Australasian	0.47%
Siberian	0.93%
Caucasus_Gedrosia	13.36%
Southern	14.23%
South_Asian	1.28%

Ethnicity calculators can also represent your estimates in a pie chart.

4. **HarappaWorld** (Harappa Ancestry Project) <www.harappadna.org> focuses on South Asian ancestry and populations: Indians, Pakistanis, Bangladeshis, and Sri Lankans. The HarappaWorld calculator has no variations.
5. **Ethio Helix** (Intra African Genome-Wide Analysis) <ethiohelix.blogspot.com> focuses on African ancestry and populations. There are four models of the Ethio Helix calculator, and the default model is *Ethio Helix K10 + French*.
6. **puntDNAL** focuses primarily on Africa (particularly East Africa), West Asia, and Europe. There are five models of the puntDNAL calculator, and the default is *puntDNAL K10 Ancient*.
7. **gedrosiaDNA** focuses on the Indian subcontinent. There are nine models of the gedorsiaDNA calculator, and the default is *Eurasia K9 ASI*.

The results of the GEDmatch analysis can be displayed in multiple ways, including a chromosome view showing where the ethnicities are found within the chromosomes and a percentages view that shows the overall percentages of the ethnicity estimate. For example, in image Ⓙ, my DNA was analyzed using the *World9* model of the Dodecad calculator. The nine reference populations used for the *World9* model are shown in the image, and the ethnicities are shown on the chromosomes. Unlike the 23andMe chromosome browser, only one copy of each chromosome is shown.

A comparison of the 23andMe ethnicity chromosome browser and the results from the *World9* model of the Dodecad calculator, shown in image Ⓚ, reveal that many segments were identified in both calculators. Generally, you can trust an ethnicity assignment that has been identified by two or more independent calculators.

In image Ⓛ, the results of the same Dodecad *World9* analysis are shown in percentage and pie chart form. Using these two formats, the test-taker can see percentages for ethnicity estimates, as well as where within the chromosomes those segments are located.

Limitations of Ethnicity Estimates

Ethnicity estimates are subject to several inherent limitations that prevent them from being completely accurate or especially helpful for genealogical research. These limitations do not mean that ethnicity estimation is bad science; rather, the limitations mean the science these estimates are based upon is continuing to develop and improve. And as a result, it's almost certain that any ethnicity estimate you receive today will be revised and updated several times in the future.

First, it's important to remember that ethnicity estimates are just that—estimates. Although the estimates have genealogical applications, they are fundamentally limited by the underlying science. For example, every company or third-party calculator utilizes a reference population, but reference populations are based on *modern-day* populations rather than ancient populations. Further, these reference populations sample a limited number of people and are not representative of the entire world.

Further, some ethnicities are nearly impossible to accurately identify. For example, populations have been migrating throughout central and western Europe for centuries, bringing their DNA from one place to another. Accordingly, the populations that eventually became modern-day Germany, France, Belgium, Switzerland, and a variety of other locations do not have enough genetic differences to be able to reliably identify a test-taker's DNA as belonging to just one of those populations. AncestryDNA describes this genetic intermingling process in its Help Topics section:

> "When individuals from two or more previously separated populations begin intermarrying, the previously distinct populations become more difficult to distinguish. This combination of multiple genetic lineages is called **admixture**. Regions that border each other are often admixed — sometimes to a great degree."

For example, AncestryDNA has found that most of the people in their Spain reference panel have about 13 percent of their DNA from the Italy/Greece region. Accordingly, it is difficult to determine whether an individual who has DNA from the Italy/Greece region has recent Italian/Greek or Spanish ancestry.

While broad categories such as Europe, Asia, Africa, and the Americas are generally reliable, ethnicity estimates become less reliable the more specific the estimate attempts to predict. Accordingly, a test-taker must be cautious about relying on an ethnicity estimate at the sub-continent or country level.

Genealogical Uses of Ethnicity Estimates

Despite their limitations, ethnicity estimates can have genealogical applications. For example, 23andMe's Ancestry Composition has a chromosome view that shows the test-taker where the segments of DNA for each ethnicity are found. The test-taker whose results are in image Ⓜ has African (red) and Native American (orange) segments of DNA on chromosome 2. Although the exact start and stop positions of these segments of DNA (which would more specifically identify shared DNA) are not provided, the test-taker can use the information to look for others who share similar segments. It may also be possible to assign these segments of DNA to particular ancestors if the test-taker knows what parent, grandparent, or ancestor these segments likely came from.

M

2

Looking for other individuals who have DNA from similar parts of the world in particular places on particular chromosomes can be a new avenue of research.

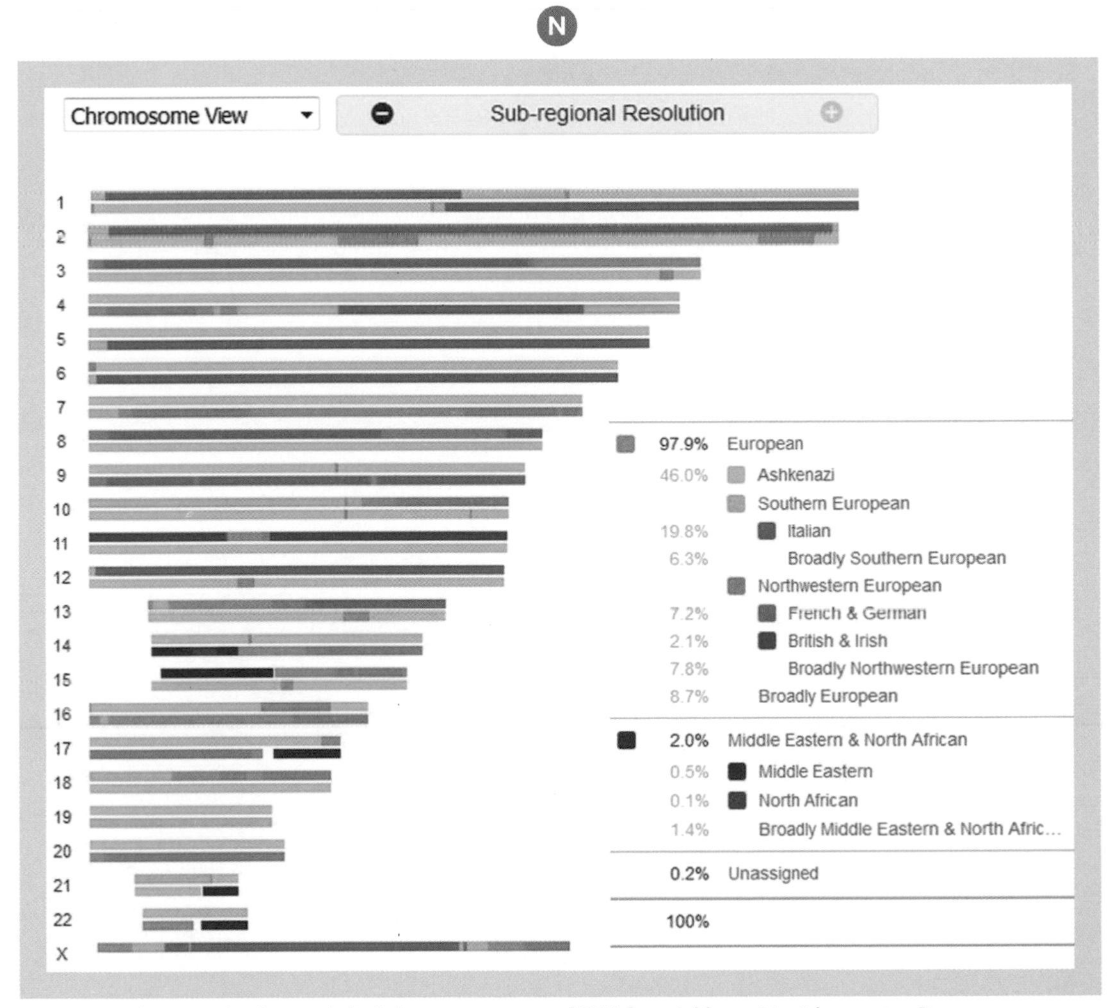

This test-taker has a relatively large percentage of DNA from Ashkenazi Jewish ancestry. Because none of the Ashkenazi regions overlap on both chromosomes, it's likely the test-taker inherited the DNA from only one parent, opening up a research opportunity.

Ethnicity estimates can also provide clues to adoptees or others with recent genealogical brick walls. In image N, for example, one of the test-taker's parents had almost 100 percent Ashkenazi Jewish ancestry. As a result, the test-taker is predicted to be about 46 percent Ashkenazi, and none of the Ashkenazi segments (in green) overlap on both copies of the chromosome. This means that the adoptee almost certainly inherited the segments from one parent.

CORE CONCEPTS: ETHNICITY ESTIMATES

- **An ethnicity estimate represents which portions (and how much) of a test-taker's DNA match one or more reference populations around the world.**

- **A reference population is a collection of DNA samples representing a particular geographic population at some recent point in time.**

- **The small size and limited geographic diversity of a reference population database constrains the accuracy of ethnicity estimates created using that database.**

- **Each of the testing companies—23andMe, AncestryDNA, and Family Tree DNA—provides an ethnicity estimate with atDNA testing. Third-party tools such as GEDmatch offer additional ethnicity calculators.**

- **Ethnicity estimates are not able to adequately distinguish between specific geographic locations such as neighboring countries. Ethnicity estimates work best for determining the continental source of DNA (Africa, Americas, Asia, and Europe).**

- **Ethnicity estimates can sometimes provide useful information to genetic genealogists as long as you bear in mind their limitations.**

Global Regions Comparison Worksheet

The accuracy of ethnicity estimates depends largely on the geographic regions each testing company uses to sample and report data. Below, you'll find a table expressing the global regions used by each of the "Big Three" (23andMe, AncestryDNA, and Family Tree DNA) in reporting ethnicity estimates as of this book's writing. Download a PDF version online at <**ftu.familytreemagazine.com/ft-guide-dna**>.

Continent	23andMe	AncestryDNA	Family Tree DNA
Africa	Middle Eastern & North African • Middle Eastern • North African Sub-Saharan African • West African • East African • Central & South African	Africa • Africa North • Africa Southeastern Bantu • Benin/Togo • Ivory Coast/Ghana • Nigeria Africa South-Central Hunter-Gatherers • Cameroon/Congo • Mali • Senegal	Africa • East Central Africa • North Africa • West Africa • South-Central Africa
America	Native American (Under Asia)	Native American	Native American
Asia	South Asian East Asian & Native American • East Asian • Korean • Japanese • Chinese • Mongolian • Yakut Southeast Asian	Asia • Asia South • Asia East • Asia Central West Asia • Middle East • Caucasus	Central/South Asian • Central Asia • South Asia • Southeast Asia East Asia • Northeast Asia Middle Eastern • Eastern Middle East Asia Minor
Europe	Northwest European • British & Irish • Scandinavian • Finnish • French & German Southern European • Sardinian • Italian • Iberian • Balkan Eastern European Ashkenazi	Great Britain Europe West Ireland Italy/Greece Scandinavia Iberian Peninsula Europe East European Jewish Finland/Northwest Russia	European • Western and Central Europe • Eastern Europe • Southern Europe • British Isles • Finland and Northern Siberia • Scandinavia • Ashkenazi Diaspora Blended Population Clusters • British Isles, Western & Central Europe • Eastern, Western & Central Europe • Scandinavia, Western & Central Europe • Southern, Western & Central Europe
Oceania	Oceanian	Pacific Islander • Melanesia • Polynesia	(none)

10

Analyzing Complex Questions with DNA

You have a brick wall in the mid-1800s that you've been trying to solve for years. (I know you do, because we all do!) These brick walls can be incredibly stubborn, and every possible source of evidence, including DNA evidence, should be considered. Today, many of these brick walls are falling due to the power of DNA.

In addition to breaking down brick walls (or at least allowing you to see over them), genetic genealogy can provide evidence for supporting or rejecting hypothesized relationships and can confirm established and well-researched lines. In this chapter, we will examine some of the ways that DNA testing can be used to provide this evidence.

Solving Questions with mtDNA and Y-DNA

Both mitochondrial-DNA (mtDNA) and Y-chromosomal (Y-DNA) testing can be powerful tools for breaking through brick walls. In chapter 5, we examined how Y-DNA can be used to analyze a paternal relationship between two or more males, and in chapter 11 we'll see several ways that Y-DNA testing can be used to help adoptees with their search. Using these techniques, Y-DNA can shed light on many of the questions asked by genealogists.

Similarly, mtDNA can be utilized with almost identical techniques to help analyze and solve complex genealogical questions.

To confirm or reject a hypothesized paternal relationship—or to verify an established and well-researched paternal line—you need to test at least two males. Only in rare circumstances can a genealogical question be answered by testing a single male. One such circumstance is when a particular ethnic ancestry is being examined. For example, a family might suspect that a great-great-great-grandfather was Native American on his paternal line. A Y-DNA test of a direct-line paternal descendant will provide evidence to support or reject this hypothesis based solely on the haplogroup. If the Y-DNA belongs to a Native American haplogroup, the hypothesis is supported. If the Y-DNA does not belong to a Native American haplogroup, the hypothesis must be updated and possibly rejected.

Another circumstance where testing only a single male might yield results is if the test-taker has a large surname project to which he can compare his results. For example, a Williams man who tries to confirm his Y-DNA line by comparing his results to the results of known paternal relatives in the Williams surname project might only have to test himself. However, if his results indicate that he is not actually related to any of the Williams test-takers in the surname project, he will undoubtedly end up testing other individuals or wait for another individual to test and provide a closer match.

Similarly, for mtDNA, you'll need to test at least two people—male or female—to confirm or reject a hypothesized maternal relationship or to verify an established and well-researched maternal line. In rare circumstances similar to those we saw with Y-DNA, including distinct ethnicities and DNA projects that incorporate mtDNA test results, mtDNA from a single person might provide sufficient information.

If Y-DNA is being utilized, the test-taker should consider a 37-marker test (or, preferably, a 67-marker test). For mtDNA, you should use a whole-genome test. In almost every case, it will be important to provide as detailed a relationship prediction as possible, and this cannot be done with low-resolution tests.

In the following example (image A), Ben Albro has hit a brick wall at his great-great-grandfather, Seth Albro. Ben has no information or clues about Seth Albro's parentage or place of origin, and he hopes that a Y-DNA test might shed some light on the mystery. Ben has ordered a 37-marker test for himself, and when he receives the results he has several close Y-DNA matches:

Genetic Distance	Name	Most Distant Ancestor
0	George Albro	Job Albro, b. 1790 Rhode Island
0	Victor Albro	Job Albro, b. 24 May 1790 Rhode Island
1	Jameson Albro	Unknown

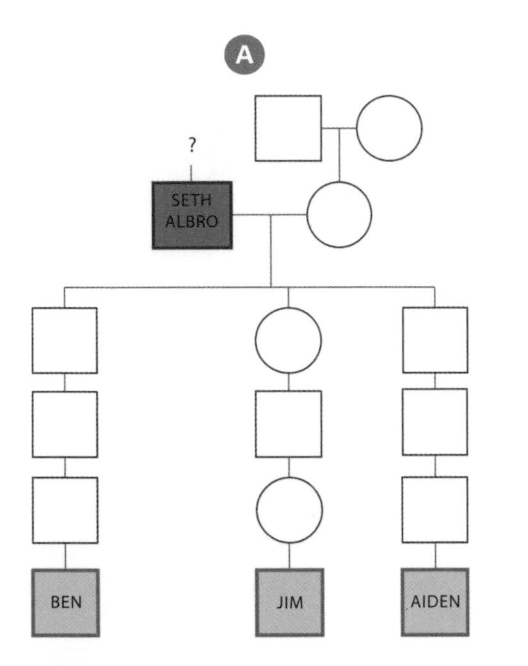

Y-DNA testing can help you find ancestors when other kinds of genealogical research fails. Here, Ben is trying to learn about his paternal great-great-grandfather's ancestors.

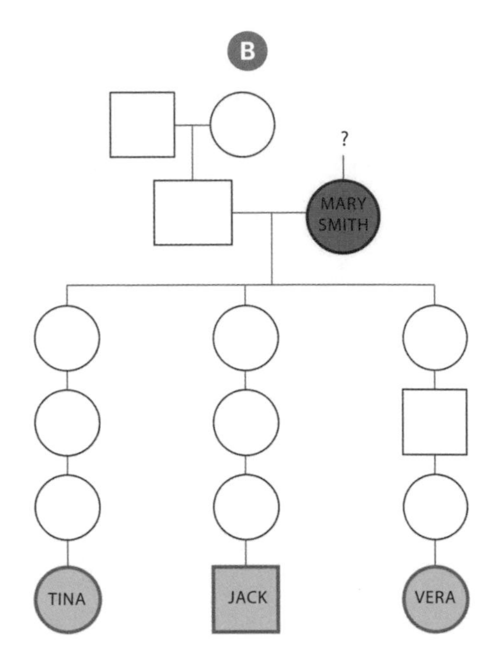

Like Y-DNA, mtDNA can be used to investigate genealogical questions. Tina can use mtDNA testing to find information about her maternal great-great-grandmother, Mary Smith.

Ben is an exact match with both George and Victor Albro, who can trace their ancestry back to an Augustus Albro, born in 1790 in Rhode Island. Ben joined the Albro surname project before ordering his test, and shortly after his test results were available, the administrators let him know that he most closely matches the Job Albro haplotype, a group of individuals who are descended from Job Albro of Rhode Island. Ben now has significant clues as to where to look for Seth's parentage, although it may not be enough to definitively identify Seth's ancestry or even his paternal line. Autosomal-DNA (atDNA) testing may provide other clues for Ben to pursue, as discussed later in this chapter.

It is important to remember as well that a brick wall is not a prerequisite for Y-DNA or mtDNA testing. Even if Seth Albro's ancestry had been well known, Y-DNA testing one or more of Ben's paternal cousins can confirm the last few generations in each of his lines. Or test results might identify an unknown break in one of these lines that could then be analyzed. For example, Seth could have asked Aiden to take a Y-DNA test for comparison. If he (Ben) and Aiden were a sufficient match, Ben could confirm that both lines go back to the paternal common ancestor Seth Albro. Cousin Jim could not, of course, provide a Y-DNA sample since he is not a direct-line male descendant of Seth Albro.

Similar to Y-DNA testing, mtDNA testing can be used for examining complex genealogical questions. One important caveat, however, is that mtDNA is not as good as Y-DNA at estimating the genealogical distance between mtDNA matches. A perfect mtDNA match, even using whole-genome matching, can mean that two people share a very recent maternal ancestor or that they share a very distant maternal ancestor. Another caveat is that there will not be a correlation between the mtDNA and the surname of either the test-taker, the genetic match, or the ancestor in question. With mtDNA—unlike with Y-DNA—the surname will most likely change at every generation.

In this example (image B), Tina has hit a brick wall at her great-great-grandmother, Mary Smith. Tina currently has no clues or other information about Mary's maiden name, her parents, or where or when she was born. Tina takes an mtDNA test with hopes that the results will shed some light on the mystery. When they come back, the results indicate that Tina has one exact mtDNA match: Ms. S. Connor, with a most distant ancestor of Jane Thompson (born about 1770 in Virginia) and the haplogroup *A2w*.

Tina can now contact Ms. S. Connor to introduce herself and ask about this line. Although it is not clear if Tina and Connor are closely related within a genealogically relevant time frame, it is potentially an important clue that Tina should pursue.

If Tina is interested in confirming her line of descent from Mary Smith, she could also ask her cousin Jack to take an mtDNA test. Although Jack is a male, he should have the same mtDNA as Tina if Tina's research is correct. If they do indeed share the same mtDNA, that would confirm both lines of descent back to Mary Smith. Cousin Vera, in contrast, is not a suitable relative to take the mtDNA test because Vera's grandfather created a break in the mtDNA line between Vera and Mary Smith.

These are just a few examples of how mtDNA and Y-DNA can be used to examine and possibly answer complex genealogical questions. To stay on top of the latest developments and learn about other ways to utilize mtDNA and Y-DNA, join surname projects, haplogroup projects, and/or geographic projects at resources like Family Tree DNA **<www.familytreedna.com>** to interact with other genealogists in forums and social media and read about success stories in works of genealogical scholarship.

Solving Complex Questions with atDNA

atDNA is perhaps the most promising new tool for analyzing complex genealogical questions. As the sizes of the atDNA databases grow, and as family tree data is combined with the results of atDNA testing, the power of DNA will continue to grow. Many questions that previously could not be analyzed—much less answered—might be easily addressed by atDNA. In this section, we will look at several different ways that atDNA can be used

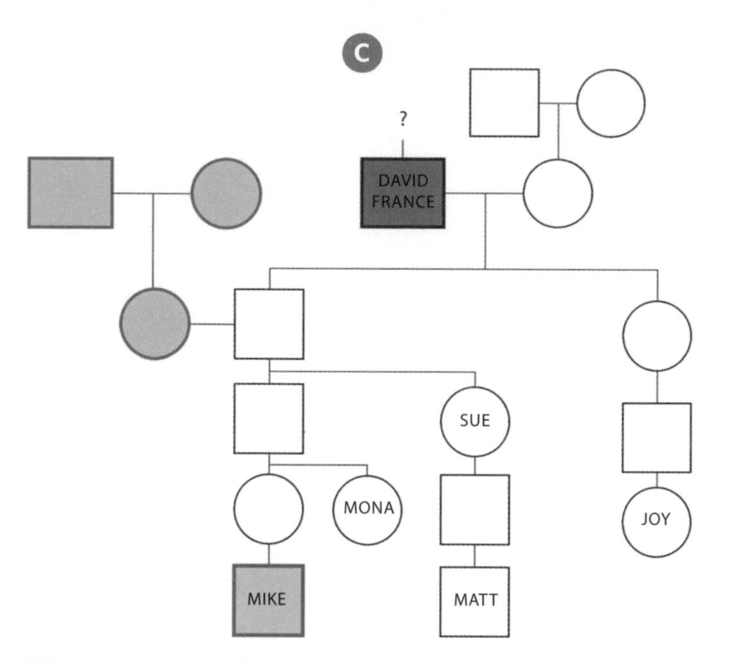

Mike can use atDNA to find David France's parents. He has many potential test-takers (such as Mona, Sue, Matt, and Joy) to choose from.

to examine genealogical questions, including both segment and tree triangulation, among other methods.

One of the key considerations in any DNA testing plan is deciding who should be tested. In a perfect world where money grows on trees, we could test every possible relative who agreed to testing. In the real world, however, we have finite resources and must be more careful about testing. Accordingly, when examining a specific genealogical question, we must first test the individual or individuals who are most likely to provide evidence about that question.

In the example displayed in image C, Mike would like to identify the parents of his ancestor, David France. Mike has taken an atDNA test at each of the three testing companies and transferred his DNA to GEDmatch <**www.gedmatch.com**>, but he has not yet uncovered any strong evidence of David France's ancestry. Mike currently has funds for one atDNA test, and he would like to know which person to ask to take the test for him. He has four potential candidates: Mona (his aunt), Sue (his great-aunt), Matt (his second cousin), and Joy (his second cousin once removed).

So who should Mike pick? Any of these relatives will potentially provide relevant information and genetic matches, but Sue and Joy are likely to be the best candidates for atDNA testing given the set of facts. Sue and Mike will only share DNA from one additional set of ancestors (indicated in blue in the image). In contrast, Mike and his aunt Mona will share many more ancestral lines in common, and it will be difficult to narrow any shared DNA or matches to just those that result from David France and his unknown parents.

Joy is a good candidate because any DNA that Mike shared in common with Joy likely came from David France and his wife (barring any other recent ancestry on their other ancestral lines). This significantly increases the importance of any shared DNA/matches by Joy and Mike, all of which could potentially be of interest and should be pursued. However, Mike will likely share less DNA with Joy than he will with Sue, although the DNA shared with Joy will potentially be of greater interest. This is a circumstance where, like so many others involving atDNA, there is no definitive answer until the tests are ordered and analyzed.

To maximize the information obtained by atDNA testing, you'll often have to test multiple descendants of an ancestor or ancestral couple. For example, if Mike eventually expands his research and identifies more descendants for testing, he will greatly increase his chance of finding genetic matches with good family trees that are related to him through David France. Mike can test Mona, Sue, and Joy (or the ancestors in blue) to obtain additional information and test results.

Not only will Mike identify segments of DNA that he shares with each of these individuals, but he will also identify segments of DNA shared by other descendants and not shared with Mike, thereby forming a genetic network of segments and shared matches that can be mined and explored. Even segments that are shared by just two of the descendants can be utilized in this genetic network research project.

Tree Triangulation

Tree triangulation is a term coined by the genetic genealogy community to refer to finding shared ancestors among the trees of close relatives and building a potential family tree connection using those shared ancestors. At its most basic form, tree triangulation typically involves the following steps:

1. Review the trees of the test-taker's closest matches to look for shared surnames and ancestors or relatives among these trees.
2. Find networks of matches with these shared surnames and ancestors using In Common With (ICW) status, such as the ICW button at Family Tree DNA and the Shared Matches feature at AncestryDNA.
3. Review the network and family trees to try to find candidates for the test-taker's ancestors, particularly parents or grandparents (if the test-taker is an adoptee).

Step one of the process will typically involve reviewing many family trees of close relatives to find patterns of surnames and ancestors. For example, if you observe that the

Philips surname is found in several of the family trees, additional research will be needed to determine whether it is the same Philips surname, which will be further supported if the genetic matches to whom those family trees belong are shared in common with the test-taker, at step two of the method.

As an example, Dianna reviews the family trees of her closest matches at AncestryDNA, including a predicted second cousin and ten predicted third cousins. The second cousin has a family tree, as do five of the third cousins. While reviewing these family trees, Dianna notices that the Pierce surname is found in both her second cousin's family tree and one of her third cousins' trees. The trees suggest that her second cousin and third cousin are themselves second cousins, and the Shared Matches tool at AncestryDNA shows that they do in fact share DNA in common. The tool also shows that they both share DNA with another third cousin that has a Worthington line that is collateral to the Pierce line in the family trees of the second and third cousins (i.e., the Worthington line married into the Pierce family). This is an incredibly strong clue that Dianna is either very closely related to or descended from this same Pierce family, although other lines will have to be considered until additional information or matches are available. Dianna can now pursue this lead and fill out the family tree of this Pierce family.

Tree triangulation is still a relatively new and unexplored methodology, and thus has not yet reached its full potential. It is expected that this methodology will continue to gain new adherents and mature as more people explore and understand the concepts underlying the process.

Segment Triangulation

One of the primary goals of atDNA research is to find a common ancestor with a genetic match, which allows you to assign the segment of DNA shared with that genetic match to the common ancestor. This process of identifying the potential source of a segment of DNA is called **segment triangulation**. Triangulation is extremely challenging and comes with many caveats, but can potentially facilitate the identification of common ancestors shared with new genetic matches.

More formally, triangulation can be defined as a technique used to identify the ancestor or ancestral couple potentially responsible for the DNA segment or segments shared by three or more descendants of that ancestor or ancestral couple. Triangulation involves the combination of DNA and traditional records in order to assign a segment of DNA to an ancestor. In chapters 6 and 8, we learned about the different chromosome browsers available from the testing companies and GEDmatch. These chromosome browsers are important sources of the information needed for triangulation.

Triangulation is a very advanced technique, and is one of the most time-consuming methodologies currently used by genetic genealogists. Accordingly, it should only be considered if the lower-hanging fruit of tree triangulation and similar methodologies have not provided the necessary information. Triangulation works by either using a third-party tool that automates the process, or by creating a spreadsheet with segment data including at least the chromosome number, start position, and stop position of each shared segment. Once the third-party tool is utilized or the spreadsheet is created, the test-taker can look for segments of DNA that are shared in common by two or more other individuals. If there are at least three people who share a segment of DNA and share a common ancestor, that is evidence—but not proof—that the segment of DNA may have come from the common ancestor.

STEP 1: DOWNLOAD SEGMENT DATA

Download segment data from each of the testing companies and/or from GEDmatch. For example, 23andMe **<www.23andme.com>** and Family Tree DNA both provide segment data for download into a spreadsheet. Family Tree DNA provides the information for all genetic matches, while 23andMe only provides the information for individuals who share genomes with the test-taker. In contrast, AncestryDNA does not share any segment data with test-takers. As a result, the only way to get segment data from matches at AncestryDNA is to ask them to upload their raw data to GEDmatch where segment data is freely available.

You can obtain your segment data from GEDmatch fairly easily by following a few steps. First, perform a One-to-many DNA comparison using the kit of interest (see chapter 8 for more details). Click the Select box for any individuals of interest in the One-to-many DNA comparison result list, then click Submit on the same page. On the next page, click Segment CSV file to obtain a spreadsheet of segments shared with the individuals of interest.

1

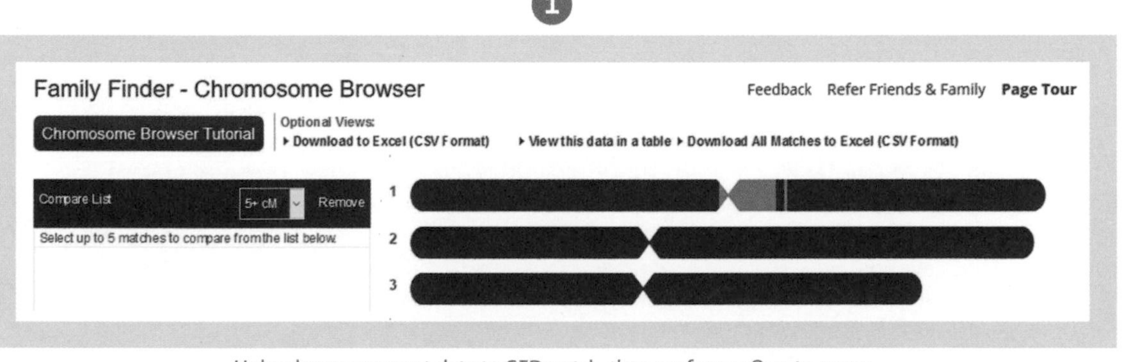

Upload your segment data to GEDmatch, then perform a One-to-many DNA comparison and download a CSV file of the segments.

STEP 2: CREATE A TRIANGULATION SPREADSHEET

Once the segment data is obtained from 23andMe, Family Tree DNA, and/or GEDmatch, it can be collated into a single master spreadsheet. You can use various column titles, such as:

- Test-taker
- Company
- Match Name
- Chromosome
- Start Location
- End Location
- Centimorgans (cMs)
- Suggested Relationship
- E-mail Address
- Surnames

Most likely, the spreadsheet is many thousands of rows long, and most of this data is small segment data. Many genetic genealogists then remove some or all of the small segments from the spreadsheet, and you'll need to find some way of sifting through all that data. For example, I always sort the spreadsheet in Excel or the spreadsheet program by

TRIANGULATION USING THIRD-PARTY TOOLS

GEDmatch and DNAGedcom <**www.dnagedcom.com**> have many tools to assist you with triangulation, and we examined some of these in chapter 8. For example, JWorks is an Excel-based tool that allows you to create sets of overlapping DNA and assign ICW status within the sets, and KWorks is similar to JWorks, except it runs in the browser. Note that the output of the two programs is based on ICW status and thus is not actual triangulation. Based solely on these analyses, you don't known for certain if the individuals share the actual identified overlapping segments in common, just that they share some DNA in common.

GEDmatch also has some other tools to help with triangulation. For example, among the Tier 1 tools is the Triangulation tool, which identifies people in the GEDmatch database who match the test-taker, then compares those matches against each other to perform true triangulation. Results can be sorted by chromosome or kit number and can be displayed in either tabular or graphical form. The results can be fine-tuned by setting a minimum (to remove small segments) and/or maximum (to eliminate close relatives) amount of DNA to be displayed.

DNAGedcom also offers the Autosomal DNA Segment Analyzer (ADSA), which is a visual ICW and triangulation tool. ADSA constructs online tables that include match and segment information as well as a visual graph of overlapping segments, together with a color-coded ICW matrix that allows pseudo-triangulation of segments.

2

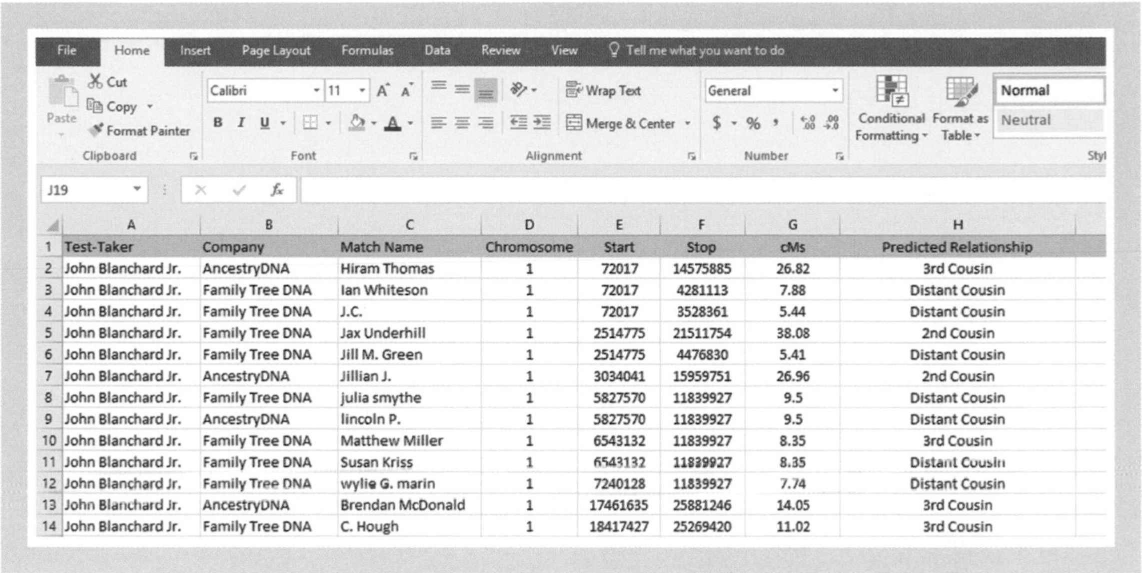

	A	B	C	D	E	F	G	H
1	Test-Taker	Company	Match Name	Chromosome	Start	Stop	cMs	Predicted Relationship
2	John Blanchard Jr.	AncestryDNA	Hiram Thomas	1	72017	14575885	26.82	3rd Cousin
3	John Blanchard Jr.	Family Tree DNA	Ian Whiteson	1	72017	4281113	7.88	Distant Cousin
4	John Blanchard Jr.	Family Tree DNA	J.C.	1	72017	3528361	5.44	Distant Cousin
5	John Blanchard Jr.	Family Tree DNA	Jax Underhill	1	2514775	21511754	38.08	2nd Cousin
6	John Blanchard Jr.	Family Tree DNA	Jill M. Green	1	2514775	4476830	5.41	Distant Cousin
7	John Blanchard Jr.	AncestryDNA	Jillian J.	1	3034041	15959751	26.96	2nd Cousin
8	John Blanchard Jr.	Family Tree DNA	julia smythe	1	5827570	11839927	9.5	Distant Cousin
9	John Blanchard Jr.	AncestryDNA	lincoln P.	1	5827570	11839927	9.5	Distant Cousin
10	John Blanchard Jr.	Family Tree DNA	Matthew Miller	1	6543132	11839927	8.35	3rd Cousin
11	John Blanchard Jr.	Family Tree DNA	Susan Kriss	1	6543132	11839927	8.35	Distant Cousin
12	John Blanchard Jr.	Family Tree DNA	wylie G. marin	1	7240128	11839927	7.74	Distant Cousin
13	John Blanchard Jr.	AncestryDNA	Brendan McDonald	1	17461635	25881246	14.05	3rd Cousin
14	John Blanchard Jr.	Family Tree DNA	C. Hough	1	18417427	25269420	11.02	3rd Cousin

GEDmatch allows you to download your segment data in spreadsheet format, allowing you to sort by a number of categories like chromosome number and length of shared segment.

the cMs column (from larger to smaller), then delete any row where the shared segment is less than 5 cMs. Don't fret if this is a significant percentage of the spreadsheet. These small segments can be problematic, and at least one study has suggested that the majority of segments of 5 cMs and smaller are in fact false positives. Once you master this process, you can go back and use the smaller segments with caution.

Once the small segments are removed, sort the spreadsheet by chromosome and start location. This will align the segments into potential triangulation groups, or groups of potentially overlapping shared segments of DNA.

STEP 3: IDENTIFY TRIANGULATION GROUPS

The goal now is to find triangulation groups, or groups of three or more people who not only share a similar segment of DNA, but are known to share that segment of DNA in common. If you don't know whether they share the segment of DNA in common, they do not form a confirmed triangulation group. Instead, they would form a pseudo-triangulation group.

In order to find triangulation groups, you will need to learn whether members of a potential group share the segment in common. This can be accomplished using, for example, the following tools:

- **23andMe:** Use the DNA comparison tool to determine whether matches share segments of DNA in common.

3

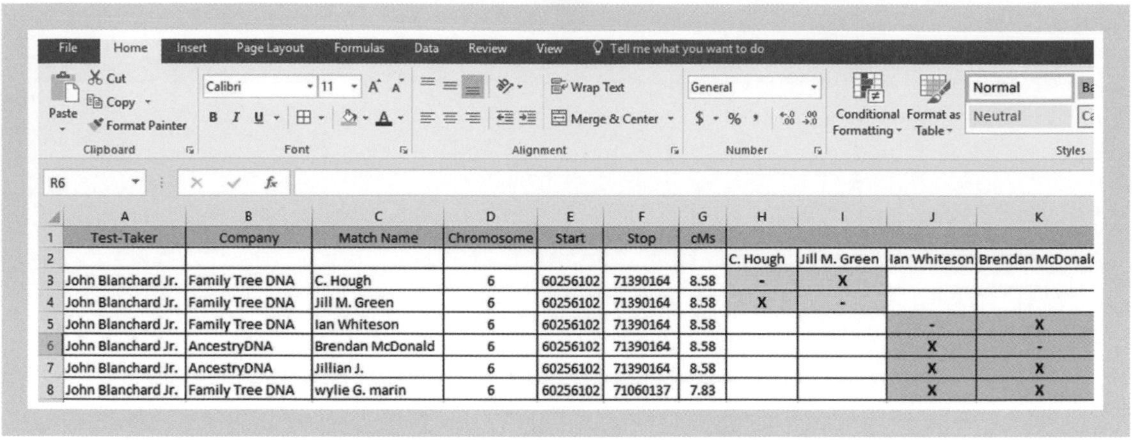

	A	B	C	D	E	F	G	H	I	J	K
1	Test-Taker	Company	Match Name	Chromosome	Start	Stop	cMs				
2								C. Hough	Jill M. Green	Ian Whiteson	Brendan McDonal
3	John Blanchard Jr.	Family Tree DNA	C. Hough	6	60256102	71390164	8.58	-	X		
4	John Blanchard Jr.	Family Tree DNA	Jill M. Green	6	60256102	71390164	8.58	X	-		
5	John Blanchard Jr.	Family Tree DNA	Ian Whiteson	6	60256102	71390164	8.58			-	X
6	John Blanchard Jr.	AncestryDNA	Brendan McDonald	6	60256102	71390164	8.58			X	-
7	John Blanchard Jr.	AncestryDNA	Jillian J.	6	60256102	71390164	8.58			X	X
8	John Blanchard Jr.	Family Tree DNA	wylie G. marin	6	60256102	71060137	7.83			X	X

Compile your segment data from testing companies and GEDmatch to analyze what segments you share with other individuals. Once you have this information, you can run an ICW analysis to see what DNA you share in common.

- **AncestryDNA:** Use the Shared Matches tool to see if two matches share DNA in common. Note that this only works for fourth cousins and closer and that sharing DNA in common indicates—but does not prove—that two people share a segment of interest in common.
- **Family Tree DNA:** Use the ICW tool. Note that overlapping segments and ICW status indicates—but does not prove—that two people share the segment of interest in common.
- **GEDmatch:** Use the One-to-one tool to determine whether matches share segments of DNA in common.

Let's look at an example. Assume that I download my segment data from Family Tree DNA, 23andMe, and GEDmatch and use that information to create a master spreadsheet. Analysis of the spreadsheets reveals that I share a segment of DNA in common with four individuals on chromosome 3, between 20 and 40 cMs in length. Without more information, I do not know whether these matches share DNA with each other, and thus whether the five of us form a single triangulation group.

If the four individuals have also tested at Family Tree DNA, I can use the ICW tool or the Matrix tool to determine which of these four individuals share *some* DNA in common with each other, although I won't know if they share the exact segment of interest in common. However, I might be able to use the information to create potential or pseudo-triangulation groups. When I use the Matrix tool, I see that the five individuals form two groupings: one group with three people who all share DNA in common and a second group with two people who share DNA in common. I, of course, am a member of both of these groups. Although I haven't confirmed using the Matrix tool that these are the actual triangulation groups, based on the results I am fairly confident that the groupings are accurate. I will also try to confirm the groupings at GEDmatch if the individuals have transferred their results to the third-party tool.

Once I have the identified triangulation groups, we can now work together as groups to compare our family trees and potentially find an ancestor in common who might be the source of the shared DNA.

Limitations

Both tree triangulation and segment triangulation are good methods to create clues for further research and add evidence to an existing hypothesis. However, neither tree triangulation nor segment triangulation is error-proof. Both methods are susceptible to a major concern, one that must be adequately addressed in any conclusion or proof argument that relies on its findings: A segment of DNA could have been inherited by another ancestor, possibly one not known to be shared by all the matching cousins.

For example, in the table in image D, a genealogist has determined the number of possible ancestors in each of the past ten generations, as well as that individual's known ancestors for that generation (where "known" means having some information about the

D

Generation Number	Shared Ancestor	Matches	Total No. of Possible Ancestors	Total No. of Identified Ancestors	Total Percentage of Identified Ancestors
1	parent	siblings	2	2	100.0
2	grandparent	1st cousins	4	4	100.0
3	great-grandparent	2nd cousins	8	8	100.0
4	2nd great-grandparent	3rd cousins	16	14	87.5
5	3rd great-grandparent	4th cousins	32	28	87.5
6	4th great-grandparent	5th cousins	64	54	84.4
7	5th great-grandparent	6th cousins	128	82	64.1
8	6th great-grandparent	7th cousins	256	124	48.4
9	7th great-grandparent	8th cousins	512	148	28.9
10	8th great-grandparents	9th cousins	1024	176	17.2

While triangulating ancestors using DNA can be effective, you likely won't be able to discover all of your direct-line ancestors through triangulation. In the example above, for example, the test-taker is able to find fewer and fewer of his genealogical ancestors as he goes farther back in time.

ancestor). The table suggests that while there is decent information about the genealogist's ancestry through the sixth generation or so, the genealogist is missing information about at least 36 percent of her family tree at the seventh generation. This poses a serious limitation on any effort to identify a shared ancestor in this generation or beyond.

In any event, genetic genealogists should recognize the possibility that DNA could be shared through other lines, and consider that possibility when reaching their conclusions.

In addition to encountering significant gaps in the family trees of members of a triangulation group, you may also face the possibility that a shared segment of DNA—particularly smaller segments—may be so common within a population that trying to narrow its source to one ancestor is problematic. For example, if segment *X* is common within a particular population, and a genealogist is descended from several different members of the population, knowing from which one of those ancestors the segment was inherited is extremely challenging.

For more information about the benefits and limitations of triangulation, including links for additional reading, see the Triangulation page at the ISOGG wiki **<www.isogg.org/wiki/Triangulation>**

Conclusions

These are just a few examples of how DNA can be utilized to examine and possible answer complex genealogical questions. This is one of the most actively studied areas of genetic genealogy, and it is likely that new methodologies, company tools, and third-party tools will find new ways to maximize the results of DNA testing.

CORE CONCEPTS: ANALYZING COMPLEX QUESTIONS WITH DNA

- **Y-DNA and mtDNA are both very useful for examining complex genealogical questions, as long as the limitations of those DNA tests are carefully considered. Likewise, atDNA is a powerful new tool for genetic genealogists analyzing genealogical questions and mysteries.**

- **In tree triangulation, researchers find shared ancestors among the trees of close relatives and build a potential family tree connection using those shared ancestors.**

- **In segment triangulation, researchers identify the ancestor or ancestral couple potentially responsible for the DNA segment or segments shared by three or more descendants.**

11

Genetic Testing for Adoptees

Every line of research reaches a brick wall; even the longest, most well-researched ancestral line eventually ends with nothing more than a question mark. For some people, the line-ending brick wall is many generations back, while for others—notably, adoptees—the brick wall is just one generation ago.

Brick walls and adoption are different names for the same challenge: finding an unknown ancestor. In this chapter, we'll examine some of the ways that genetic genealogy can be used to analyze recent ancestors who are brick walls due to a variety of reasons, including misattributed parentage, adoption, donor conception, abandonment, hospital baby switches, and amnesia, among others. Due largely to the immense and sudden growth of the testing company databases, these recent brick walls are being broken down faster and easier than ever before.

Although the chapter uses the word "adoptee," it is meant to encompass the many diverse situations that can result in a very recent brick wall (usually parent or grandparent). The methodologies used to analyze these diverse situations are usually very similar. Additionally, although numerous ethical issues can be raised when using DNA testing to break through very recent brick walls, many of these issues were discussed in chapter 3 and thus won't be repeated here.

Breaking Through Brick Walls with Y-DNA

Y-chromosomal DNA (Y-DNA) testing can be a powerful tool for breaking through brick walls. In chapter 5, we examined how Y-DNA can be utilized to confirm or reject a paternal relationship between two or more males, and surname projects sometimes do this for the hundreds of men in the project.

Accordingly, Y-DNA can be used by adoptees to identify close paternal matches, some of whom might be close enough matches to provide a potential surname for the adoptee. Often, the results of Y-DNA testing will also be helpful when trying to confirm relationships that have been identified using autosomal-DNA (atDNA) testing. At Family Tree DNA's 7th International Conference on Genetic Genealogy in 2011, Family Tree DNA managing partner and COO Max Blankfeld estimated that 30 to 40 percent of male adoptees who test their Y-DNA at Family Tree DNA find clues to their biological surname **<www.yourgeneticgenealogist.com/2011/11/family-tree-dnas-7th-international_09.html>**. This figure alone demonstrates why adoptees should *always* consider Y-DNA testing to examine their heritage. (Of course, since Y-DNA testing is limited to males, female adoptees are unable to take advantage of this DNA test.)

If Y-DNA is being utilized, the adoptee should consider a 37-marker test and possibly a 67-marker test. It will be important to provide as detailed a relationship prediction as possible, and this cannot be done with low-resolution tests. Adoptees should consider joining the Adopted DNA Project at Family Tree DNA **<www.familytreedna.com/groups/adopted>** to access a community of other adoptees and excellent project administrators.

In the following example, adoptee Nathan Vaughn has performed extensive genealogical research but has not found any accessible records that fill in his family tree. Nathan learns about genetic genealogy and decides to order a 37-marker Y-DNA test from Family Tree DNA. When he receives his results, he also receives a list of genetic matches.

Genetic Distance	Name	Most Distant Ancestor	Y-DNA Haplogroup	Terminal SNP
0	Wilhelm Davidson	Henry Davidson, b. 1790 Va	R-L1	
0	Liam Davidson	Henry Davidson, b. 1790 Va	R-L1	
1	James Davidson	Donald Davidson, b. 1773 Va	R-P25	P25
2	Philip Farah		R-L1	

Nathan has two perfect, 37-of-37 marker matches (Wilhelm Davidson and Liam Davidson), and according to Family Tree DNA's calculations, there is a 95-percent probability that the most common recent ancestor was not more than seven generations ago. Nathan is closely paternally related to these two matches, and it is very likely that his biological father had the surname Davidson, although there may have been another break in the line upstream of Nathan's biological father. Nathan should consider joining the Davidson Surname Project, if such a project exists.

Nathan and James Davidson have a genetic distance of 1 (36-of-37 markers), and their relationship is potentially a little more distant than the 37-for-37 matches. Nathan and Philip Farah have a genetic distance of 2, which means either that one of their markers is different by one mutations with a value of 2 or two mutations with a value of 1. This relationship is even more distant, but could still be within a genealogically relevant time frame.

Y-DNA Success Stories

There are many examples of adoptees and genetic genealogists who have broken through their brick walls using Y-DNA. Below are just a few examples of these success stories, all of which have been peer-reviewed and published. Reading these articles will help you understand how others apply Y-DNA testing to their brick walls. Also be sure to review the success stories collected by the International Society of Genetic Genealogy (ISOGG) at <**www.isogg.org/wiki/Success_stories**>.

- Morna Lahnice Hollister, "Goggins and Goggans of South Carolina: DNA Helps Document the Basis of an Emancipated Family's Surname," *National Genealogical Society Quarterly* 102 (September 2014): 165–176. Hollister uses Y-DNA to document the basis of an emancipated family's surname.
- Warren C. Pratt, Ph.D., "Finding the Father of Henry Pratt of Southeastern Kentucky," *National Genealogical Society Quarterly* 100 (June 2012): 85–103. Pratt combines traditional records and Y-DNA testing to identify the father of an ancestor born in 1809.
- Judy Kellar Fox, "Documents and DNA Identify a Little-Known Lee Family in Virginia," *National Genealogical Society Quarterly* 99 (June 2011): 85–96. Fox uses Y-DNA testing and traditional genealogical records to verify the parentage of an ancestor born in the mid-1700s.
- Randy Majors, "The Man Who Wasn't John Charles Brown?," randymajors.com (31 December 2010) <**www.randymajors.com/2010/12/man-who-wasnt-john-charles-brown.html**>. DNA testing confirms the suspected surname of an ancestor who changed his name.

Breaking Through Brick Walls with mtDNA

Unfortunately, mitochondrial-DNA (mtDNA) testing is often not as helpful in adoption or brick wall cases, although it has solved many different genealogical mysteries. Further, leave no stone unturned in challenging cases. For example, mtDNA testing can sometimes provide hints to the ethnic or ancestral background of an unknown mother. When considering an mtDNA test, individuals should always purchase a full mtDNA sequence, which is the highest resolution test and covers all 16,569 locations on the mitochondria.

In the following example, adoptee Juniper Saunders has purchased a full mtDNA test from Family Tree DNA, and a few weeks after sending away her collection kit, she receives the following list of genetic matches:

Genetic Distance	Name	Most Distant Ancestor	mtDNA Haplogroup
0	Jennie Banks	Nancy Collins, b. 1775 (N.Y.)	H1
0	Caren West	Nancy Collins, b. 1775	H1
1	Victor Johns		H1
2	Cynthia Nunez	Nancy (Smith) Collins, b. ~1770	H1

Although the potential ancestor or maternal relative Nancy Collins was born almost 250 years ago, Juniper can communicate with these matches to request information, and can build a family tree for Nancy Collins. Although Juniper may only be related to Nancy Collins rather than descended from her, this new clue should be pursued.

Additionally, Juniper can combine this result with other DNA testing to identify other connections. For example, Juniper should investigate whether she shares atDNA with any of her mtDNA matches. She should also search her atDNA matches for individuals that might connect to the family of Nancy Collins.

mtDNA Success Stories

Although there are not as many success stories for mtDNA as there are for Y-DNA, many genetic genealogists have broken through otherwise impossible brick walls using mtDNA testing. In addition to the success stories collected by the ISOGG at <**www.isogg.org/wiki/Success_stories**>, here is one recently published article that successfully uses mtDNA testing:

- Elizabeth Shown Mills, "Testing the FAN Principle Against DNA: Zilphy (Watts) Price Cooksey Cooksey of Georgia and Mississippi," *National Genealogical Society Quarterly* 102 (June 2014): 129–52. Mills uses mtDNA testing combined with two traditional genealogical research methods to solve an ancestry mystery.

Breaking Through Brick Walls with atDNA

atDNA has the potential to solve innumerable adoption cases and break through genealogical brick walls. In addition to containing information about many different family lines—especially recent ones—it can also reveal information about the ethnicity of the test-taker's recent ancestry.

"Adoption angels"—individuals who donate their time and expertise to help adoptees—specializing in DNA solve adoption and other family mysteries every single day using the results of atDNA testing. The size of the testing company databases is making this process easier and easier by helping adoptees find closer relatives. For example, it is generally much easier to solve an adoption case with a single half-sibling or first cousin than it is with a handful of third cousins (although having just the third-cousin matches also doesn't mean the search is hopeless). As the databases continue to grow, the likelihood of finding a parent, half-sibling, aunt/uncle, or other close relationship increases considerably.

To maximize the likelihood of finding one of these close matches, it is vital that adoptees "fish in all three ponds," according to DNA expert CeCe Moore of *Your Genetic Genealogist*, meaning that an atDNA test is taken at 23andMe **<www.23andme.com>**, AncestryDNA **<www.dna.ancestry.com>**, and Family Tree DNA **<www.familytreedna.com>**. Although there is considerable overlap among the three databases, each company has test-takers who are found only in that company's database.

If the test-taker hasn't tested at all three companies, the raw data should be transferred to GEDmatch **<www.gedmatch.com>** if there aren't any significant privacy issues involved and if the test-taker is comfortable with that level of information sharing. Since GEDmatch comprises tens of thousands of test-takers from each of the three testing companies, it is the largest database of cross-company comparisons and can thus be a very important source of genetic matches. If the test-taker has tested at all three companies, GEDmatch won't be as useful; any sufficiently close matches likely to assist the adoptee will be identified by the testing company.

Finding a Close Relative

When the test-taker finds a close family member in a list of genetic matches at a testing company (such as a parent, half-sibling, aunt/uncle, or first cousin), only a minimal amount of additional research will be necessary. Most commonly, the research will determine on which side of their respective families the two individuals are related. For example, do half-siblings share a mother or a father? Is a predicted aunt or uncle on the test-taker's maternal or paternal side of the family? There are clues that can be helpful, including X-chromosome matching (chapter 7), Y-DNA matching (chapter 5), mtDNA matching (chapter 4), or ethnicity estimates (chapter 9). For example, an adopted test-taker who is 50-percent Jewish and has a traditionally Jewish Y-DNA signature should determine which side of the family a match who is Jewish obtained his Jewish ancestry from, which can help identify their relationship.

ADOPTION RESOURCES

Many resources are available to adoptees looking for more information about how to use genetic genealogy for DNA testing. Each of these resources is highly recommended and provides slightly different information. First and foremost, it is important that adoptees network and engage with other adoptees whenever possible to ensure that they are taking the right steps in their DNA testing plan, and that they are using the most recent methodologies and resources in their research.

Websites

- **DNAAdoption** <www.dnaadoption.com>: Home of the Methodology and related classes
- **Adoption and DNA** <www.adopteddna.com>: Adoption tips and success stories
- **The DNA Detectives Facebook page** <www.facebook.com/groups/DNADetectives>: The largest community of adoptees and adoption/search angels using DNA
- **DNA Testing Advisor** <www.dna-testing-advisor.com>: Testing advice from Richard Hill, an adoptee who found his biological family in 2007 using DNA (also see his terrific and engaging book, *Finding Family: My Search for Roots and the Secrets in My DNA*)

Mailing Lists and Forums

The following mailing lists and forums provide excellent DNA testing advice and have hundreds or thousands of people willing to answer questions. Some of these groups are for general testing information, while others are specific to adoption searches.

- **DNAAdoption Yahoo Group** <groups.yahoo.com/group/DNAAdoption>
- **ISOGG DNA-NEWBIE List** <groups.yahoo.com/neo/groups/DNA-NEWBIE/info>
- **Rootsweb Genealogy-DNA Mailing List** <lists.rootsweb.ancestry.com/index/other/DNA/GENEALOGY-DNA.html>
- **Unknown Fathers DNA Yahoo Group** <groups.yahoo.com/group/UnknownFathersDNA>

The additional research needed to determine the exact relationship of a predicted close relative can be as simple as reviewing a public family tree of the match, particularly if the adoptee test-taker has one or more minimal clues to his ancestry. For example, if the adoptee has a location, profession, ethnicity, or date of birth for one or both of his biological parents, he can review the trees to see where that information might align.

Although we're foregoing an in-depth discussion of the ethical issues associated with adoption for the moment (see chapter 3 for more on ethics and genetic genealogy), it is important that both adoptees and test-takers consider the ramifications of establishing contact and do it in the best way possible. Many half-siblings, aunt/uncles, or first cousins will have absolutely no idea that the adopted test-taker exists, and thus communication should consider this possibility. Although I personally believe that every individual has a fundamental and inalienable right to their genetic heritage, I understand that does not translate to a fundamental and inalienable right to a relationship with that genetic heritage.

Finding a Distant Relative

Working with more distant matches (from second cousins outward) will be more challenging, and much of the work performed by adoption angels and the adoption community is with these more distant matches. Although not impossible, working with a handful of third-cousin matches will usually be difficult and time-intensive. If the only matches an adoptee finds are fourth cousins and more distant, finding the proper family is going to be extremely difficult and potentially not possible. In this latter situation, the test-taker can work with the available matches while waiting for newer, better matches to test.

Without a doubt, all possible clues should be reviewed and pursued. At AncestryDNA, for example, review DNA Circles and New Ancestor Discoveries for clues. Although the individuals identified in a New Ancestor Discovery are most often from collateral lines or are only loosely relative, a find there will actually be an ancestor in some cases. That possibility should be considered, although the test-taker should always remember that the greater likelihood is that the New Ancestor Discovery is most often a collateral line or relative.

Using Triangulation

Both segment triangulation and tree triangulation are useful for adoptees (and all test-takers) with more distant cousin relationships, although these often help those with closer relationships as well.

As discussed in chapter 10, tree triangulation is the term for finding shared ancestors among the trees of close relatives and building a potential family tree connection using those shared ancestors. For people with known ancestry, tree triangulation is often based on looking for surnames or places that have a link to their known pedigree. For example, someone with known or suspected ancestry in southern California may logically focus on clusters of relatives with overlapping trees in southern California. Or someone with known or suspected Gilmore ancestry will logically focus on clusters of relatives with overlapping Gilmore ancestors.

Adoptees, foundlings, and others with little or no known ancestry must approach tree triangulation differently. Instead of focusing on places or surnames that have some link to a researched pedigree, adoptees must focus entirely on patterns within the trees of their close matches.

The Shared Matches tool at Ancestry DNA and the ICW tool at Family Tree DNA are excellent ways to identify patterns. For every match predicted to be a third cousin or closer, use the ICW tool to identify clusters of matches. Potentially, some of the matches within a cluster will be related through the same family, and will have at least a portion of that family in their associated family trees.

For example, assume that an adoptee named Zula has a predicted second cousin match at AncestryDNA named *J.T.2016*, and she uses the Shared Matches tool to identify matches that Zula and J.T.2016 have in common. They have four matches in common, two of whom have a family tree linked to their DNA results. When Zula reviews the family trees of J.T.2016 and their two shared matches, she sees the Westmiller surname in all three trees. In two of the trees she finds Abraham Westmiller born in 1876 in Vermont. J.T. 2016 is descended from one son of Abraham, and one of the shared matches is descended from another son of Abraham. The other shared match doesn't have Abraham Westmiller in their tree, but Zula builds out that person's tree and discovers that they are also a descendant of Abraham Westmiller.

Based on this information, Zula should build a family tree around Abraham Westmiller and his wife, including both descendants and ancestors. She can also then link her own DNA results to individuals within the constructed Westmiller family tree to see if she gets DNA hints. She should also search for matches at each of the testing companies with the Westmiller surname in their profiles or family trees.

This might be the breakthrough that Zula needs to solve her genealogical mysteries. Alternatively, Zula might have to repeat this process many times before she finds a family link that is reliable. As more people test, existing tools mature, and new tools are developed, tree triangulation should prove to be a very helpful tool for adoptees.

Adoption and The Methodology

The adoption community has developed an extensive methodology for using matches to find biological ancestors. Although designed for adoptees, the "Methodology," as it is known, can also be utilized for more distant brick walls. The Methodology was created by the DNAAdoption community, an important resource for all adoptees, and generally involves the following steps:

1. **Triangulate segments of DNA for matches that have a family tree.** The process of triangulation is outlined in great detail in chapter 10 (and addressed earlier in this chapter). Essentially, a spreadsheet of DNA segments shared with genetic matches (image A) is created from one or more testing companies or third-party tools, and that spreadsheet is used to identify triangulation groups, a collection of three or more people that all share a segment of DNA in common. If members of the identified triangulation groups have family trees available, the trees can be mined to find one or more individuals found in each of the trees. These individuals found in all three trees from the triangulation group are now candidates for the adoptee test-taker's ancestor.
2. **Create a master tree with triangulated ancestors.** Once a candidate is identified, construct a family tree around that candidate. For example, the adoptee may build a family tree forward and/or backward from that candidate. Then, he can link his DNA results to a constructed tree at AncestryDNA to view possible hints.

A

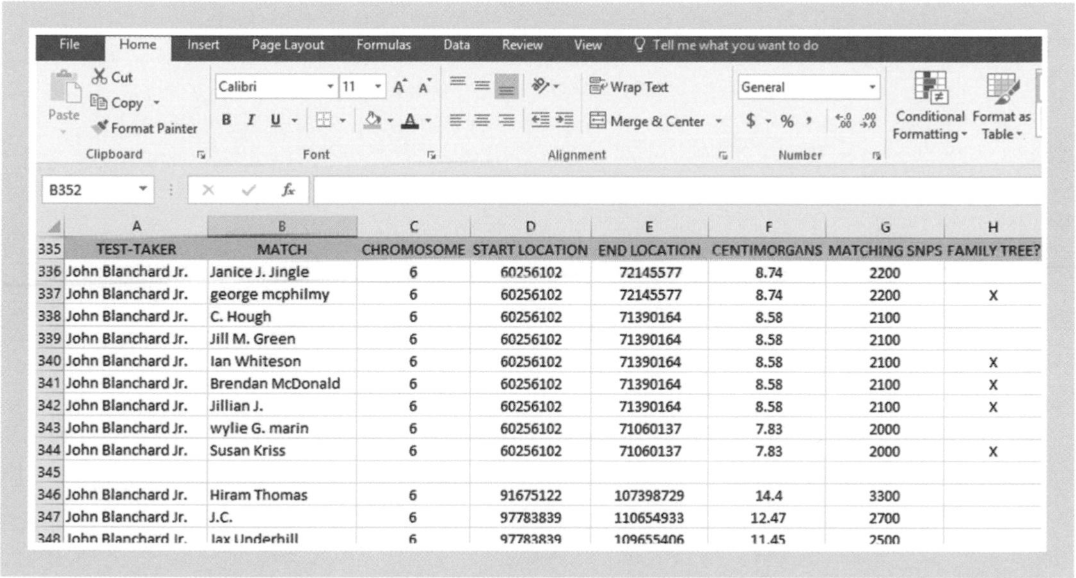

	A	B	C	D	E	F	G	H
335	TEST-TAKER	MATCH	CHROMOSOME	START LOCATION	END LOCATION	CENTIMORGANS	MATCHING SNPS	FAMILY TREE?
336	John Blanchard Jr.	Janice J. Jingle	6	60256102	72145577	8.74	2200	
337	John Blanchard Jr.	george mcphilmy	6	60256102	72145577	8.74	2200	X
338	John Blanchard Jr.	C. Hough	6	60256102	71390164	8.58	2100	
339	John Blanchard Jr.	Jill M. Green	6	60256102	71390164	8.58	2100	
340	John Blanchard Jr.	Ian Whiteson	6	60256102	71390164	8.58	2100	X
341	John Blanchard Jr.	Brendan McDonald	6	60256102	71390164	8.58	2100	X
342	John Blanchard Jr.	Jillian J.	6	60256102	71390164	8.58	2100	X
343	John Blanchard Jr.	wylie G. marin	6	60256102	71060137	7.83	2000	
344	John Blanchard Jr.	Susan Kriss	6	60256102	71060137	7.83	2000	X
345								
346	John Blanchard Jr.	Hiram Thomas	6	91675122	107398729	14.4	3300	
347	John Blanchard Jr.	J.C.	6	97783839	110654933	12.47	2700	
348	John Blanchard Jr.	Jax Underhill	6	97783839	109655406	11.45	2500	

The Methodology begins by analyzing triangulation groups, collections of individuals who share DNA (and thus, ancestors) with each other.

3. **Repeat with new matches until the brick wall is broken down.** The master tree can contain trees for every candidate identified in a multitude of triangulation groups, and eventually the tree may reveal overlapping "branches" that effectively point to a biological parent, grandparent, or other close relative.

This is only the briefest introduction to the Methodology, and each of the above steps involves several additional steps. For more information and for courses specifically tailored to adoptees, visit the DNAAdoption group's website <**www.dnaadoption.com**>.

CORE CONCEPTS: GENETIC TESTING FOR ADOPTEES

- **DNA is helping break through very recent brick walls to help adoptees, foundlings, and others identify their genetic heritage.**

- **Y-DNA testing at Family Tree DNA is able to provide a potential biological surname in as many as 30 percent of cases.**

- **mtDNA testing is not as useful for adoptees, but in rare cases can provide helpful information.**

- **The size of the testing companies' atDNA databases has greatly increased the chance that an adoptee will find a second cousin or closer in their match lists.**

- **Adoptees and others interested in using DNA to analyze their genetic heritage should consider: taking a 37- or 67-marker Y-DNA test at Family Tree DNA (if male); taking an atDNA test from *at least one* (and maybe all three) of the big testing companies; joining several adoptee-focused social media groups or mailing lists to begin to learn how to interpret and apply the Y-DNA and atDNA test results.**

12 The Future of Genetic Genealogy

As with computers and the Internet, DNA has become an important component of modern genealogical research. Although this tool only became available a few short years ago, it is now a source of evidence for thousands of genealogists and of fascination and excitement for millions of test-takers. Over the course of the next ten to twenty years, new advancements in DNA technology and techniques will change and expand how genealogists obtain DNA test results and how those test results are applied to genealogical questions. In this chapter, we will peer into our crystal ball to look at current trends in DNA technology and see how they will affect genetic genealogy.

The Future of Y-DNA Testing

One area of genetic genealogy that is likely to change considerably over the next decade is Y-chromosomal (Y-DNA) testing. Currently, Y-DNA testing comprises either a handful of Y-STRs (short tandem repeats on the Y chromosome), usually between 37 and 111, or a few thousand Y-SNPs (single nucleotide polymorphisms on the Y chromosome). Some Y-DNA tests, such as the Big Y test from Family Tree DNA **<www.familytreedna.com>**, examines approximately 15 to 25 million base pairs of the Y chromosome, which is just 25 to 45 percent of the entire Y chromosome, much of which potentially contains information about ancestry, although these tests are only just beginning to yielding information that will be helpful to genealogists.

While future researchers would ideally sequence the entire Y chromosome and analyze it for ancestral information (including identifying and characterizing new STRs and SNPs), the Y chromosome presents some unique challenges that prevent affordable or accurate whole Y-DNA sequencing with current technology. For example, much of the Y-chromosome is either highly repetitive, palindromic, or nearly identical to the X chromosome. Since current DNA sequencing technology sequences many short overlapping fragments (called "reads") of a chromosome and then pieces them back together by mapping them to a human genome reference, repetitive or palindromic sequences can make this difficult (if not impossible).

New sequencing technology, however, might provide testing companies like Family Tree DNA with new opportunities. For example, DNA sequencers that obtain very long high-quality reads will be much better equipped to piece those long reads together. Ideally, sequencing technology in the future might start at one end of a chromosome and, with one single read, sequence all the way to the other end of the chromosome.

Once raw data is available, analyzing the data to extract the STR and SNP information is an easy step in the process. Just as the influx of Big Y results from Family Tree DNA in 2015 resulted in the so-called "SNP tsunami," the data flowing in from future sequencing technology will result in a massive amount of information that will have to be analyzed and categorized in the context of the human Y-DNA family tree. There will likely be numerous "family-specific SNPs" within these results—Y-SNP variations that are found among men who have a recent shared paternal ancestor (within about the past 100 to 250 years).

There is a strong economic pressure to create improved DNA sequencing technology, including the desire to use low-cost DNA testing for health assessment and treatment. As a result (despite some technical setbacks), Y-DNA sequencing developments will likely occur within the next five to ten years.

The Future of mtDNA Testing

Since the entire mtDNA molecule is already sequenced by current testing, there will likely be only a few developments in mtDNA testing in the future. It is not possible to extract any additional information from the sequence of *A*, *T*, *C*, and *G*'s in the mitochondrial genome.

Rather, the biggest development in mtDNA testing will likely come from a greater number of people testing, meaning that the likelihood of finding a meaningful match will increase. Although many people find matches when they test and for the reasons we learned in the mtDNA chapter (namely that mtDNA mutates very slowly and thus many, many people have the same mtDNA), it is very rare that a test-taker finds a meaningful match that helps them with their genealogical research. Additionally, mtDNA is often associated with a shorter pedigree because of the challenges of researching a maternal line with a name change every generation. As the database gets larger however, the likelihood of finding an important close match that fits within your pedigree increases significantly.

Another important development in mtDNA testing might be epigenetic testing. Like atDNA, mtDNA is packaged into an organized structure with proteins and chemical groups that associate with it. If this epigenetic structure is heritable, as recent studies suggest it is, then it could be analyzed and exploited for genealogical purposes. People who are more closely related would be expected to have more similar epigenetic structure, and thus the epigenetic structure of mtDNA might single out close relatives among a list of mtDNA matches. Epigenetic testing will be explained in much greater detail later in this chapter.

The Future of atDNA Testing

The biggest changes in genetic genealogy are expected to occur in the area of autosomal-DNA (atDNA) testing, for some of the same reasons as Y-DNA testing. Namely, new developments in DNA sequencing will both improve sequencing and lower the cost of testing.

When whole-genome sequencing reaches a price point similar to current atDNA testing, genealogists will be a driving force behind the use of whole-genome results for genealogy, including improved cousin identification and relationship estimation. Whole-genome testing will provide some benefits to cousin identification and relationship estimation, although it likely won't identify new close cousins (closer than fourth cousin, for example) who couldn't be identified with current testing. In a similar way, an affordable whole-genome test for genealogists likely won't provide relationship predictions that are stunningly accurate, instead providing new distant cousins and improving the confidence in relationship estimates.

In addition to whole-genome sequencing, new atDNA methodologies will be created in the next decade. These methodologies will not only be the result of research and experimentation by scientists and genetic genealogists, but will also be possible due to the sheer sizes of atDNA databases. Once the databases comprise many millions of people, new tools can be developed that were either not apparent when the database was smaller or was not possible using a smaller database.

Genetic Reconstruction: Piecing Together Genomes of the Dead

Piecing together a portion or all of the genomes of our ancestors will enable genealogists to learn things about them that we might otherwise not be able to, such as their ethnicity, health, and recent genealogical relationships. Genealogists might also learn about some of their physical characteristics like eye color and hair color, although these are not always a perfect estimate based on DNA alone. In this section, we'll discuss what genetic reconstruction is, how it works, and why it may be of interest to genealogists.

Genetic reconstruction is made possible by testing many descendants of an ancestor or ancestral couple. For example, imagine John and Jane Smith, living in New England in the mid-1700s. They had twelve children, ten of whom lived to adulthood, and thus they now have many, many thousands of descendants living today. A handful of these descendants possess random segments of DNA handed down from John and Jane; the more children and descendants the ancestral couple had, the more DNA from that couple that is likely to exist in test-takers today.

Segments of DNA that potentially came from this ancestor or ancestral couple can be identified using well-researched family trees, then woven together to create as much of the couple's genome as possible. Some segments will be lost forever, although often these missing segments can be derived or estimated.

In image A, segments of DNA from a couple in the mid-1700s are found in living descendants. Some descendants, such as #4, either did not inherit any DNA from the couple or does not share any of the inherited DNA with another relative, and thus will likely not contribute to the genetic reconstruction. Others descendants—or relatives—such as descendant #6 may have no documentation that they are related to the couple. Only segments that can be reliably assigned to the ancestor or ancestral couple will be mapped. Usually, this will involve identifying segments of DNA shared by two or more descendants of the ancestor or ancestral couple.

Slowly, the genomes of hundreds and possibly even thousands of early ancestors may be generated as millions of DNA samples and family trees are entered into massive databases. There will undoubtedly be numerous errors introduced by both poor quality trees—or trees that are incorrect due to misattributed parentage events—and from improperly assigning shared segments to one ancestor versus another. However, most of these errors will be resolved over time as more samples and trees are entered into the system and processed and as genealogists and citizen scientists pour over the trees.

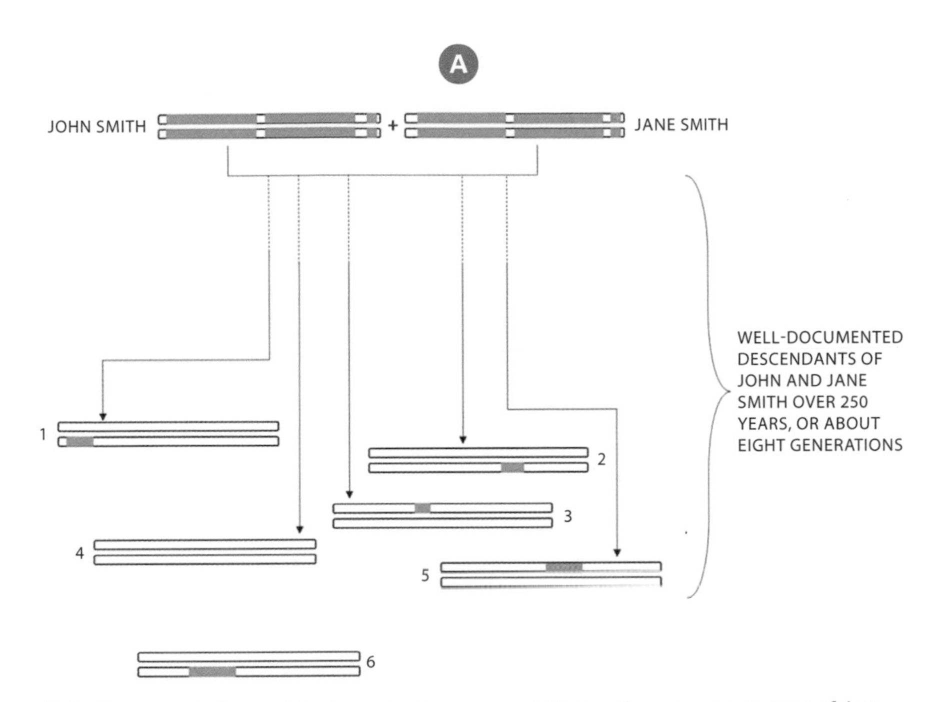

By looking several of a couple's descendants, you can cobble together an approximation of their DNA makeup. However, only some of these descendants will contain the original couple's DNA; here, descendant #4 doesn't have this ancestral DNA and so will not be helpful for these purposes.

Interestingly, these recreated genomes sometimes will belong to unknown or unidentified ancestors (DNA-Only Ancestors), as discussed later. For example, shared segments of DNA may appear to be from a couple who probably lived in the area of Boston and had children in the early 1700s, but no known couple can be found in existing records.

To be successful, this future methodology will of course require numerous, wide-ranging, and extremely well-researched family trees, as well as DNA samples from several millions of individuals.

With reconstructed genomes, it might also be possible to estimate what our ancestors looked like, even if no picture of that ancestor was ever taken or has survived. Everyone knows the age-old game of guessing which parent or sibling a new child looks like, or guessing which identical twin is which. These scenarios demonstrate the existence of a relationship between DNA and appearance. Accordingly, by examining and understanding this relationship, we can theoretically predict appearance based on DNA alone.

For example, in 2014, scientists published a study that identified twenty-four gene variants across twenty genes that affect facial structure **<journals.plos.org/plosgenetics/article?id=10.1371/journal.pgen.1004224>**. The researchers then used DNA profiles from volunteers to create approximations of the volunteer's facial structure. In addition to having numerous potential forensic and law-enforcement applications, this technology could assist genealogists who are recreating the appearance of long-dead ancestors. Facial structure estimates could be combined with other physical information mined from the DNA sequence, including eye color, hair color, height, skin color, and other physical characteristics, to create a composite image of the ancestor. The DNA information could also be supplemented with cultural and socioeconomic information to predict hair styles and other features.

In another example, AncestryDNA <**www.dna.ancestry.com**> announced in 2014 that it had successfully recreated significant fragments of the genome of David Speegle (1806–1890) and his two wives, Winifred Cranford and Nancy Garren. This was accomplished by analyzing the DNA of hundreds of Speegle's descendants through his twenty-six children and piecing together shared segments of DNA using two different methods. With many children between the two marriages during his lifetime, David and his spouses were excellent candidates for reconstruction given the number of living descendants who all potentially carry a piece of their DNA. Indeed, according to Speegle's obituary in 1890, he had at least three hundred descendants at the time of his death, suggesting why DNA from David Speegle and his wives was so prevalent in the AncestryDNA database.

Using these recreated partial genomes, AncestryDNA analysts learned that David or one of his wives had a gene variant that increases the likelihood of male pattern baldness, and that David had at least one copy of the gene variant for blue eyes. See <**blogs.ancestry.com/techroots/ancestrydna-achieves-scientific-advancement-in-human-genome-reconstruction/**> for more information.

Generating Family Trees

So what could atDNA advancements do for your documented family research? In theory, once the genomes of hundreds or thousands of seventeenth-, eighteenth-, or nineteenth-century ancestors are created and collated into a massive family tree, they can be used to recreate portions of the family tree of modern-DNA test-takers using just the results of a DNA test. This is done by first identifying potential ancestors based on the results of an atDNA test, then by fitting those identified ancestors into a family tree for the test-taker. This might be augmented, for example, by any known genealogy for the test-taker.

Family tree prediction or reconstruction is possible because identified ancestors will only fit together into a family tree in a limited number of ways. For example, let's assume you've taken an atDNA test and the testing company has identified twenty ancestors using only its database of reconstructed ancestors and your DNA test results. Statistically speaking, there are a limited number of ways to collate those twenty ancestors into a

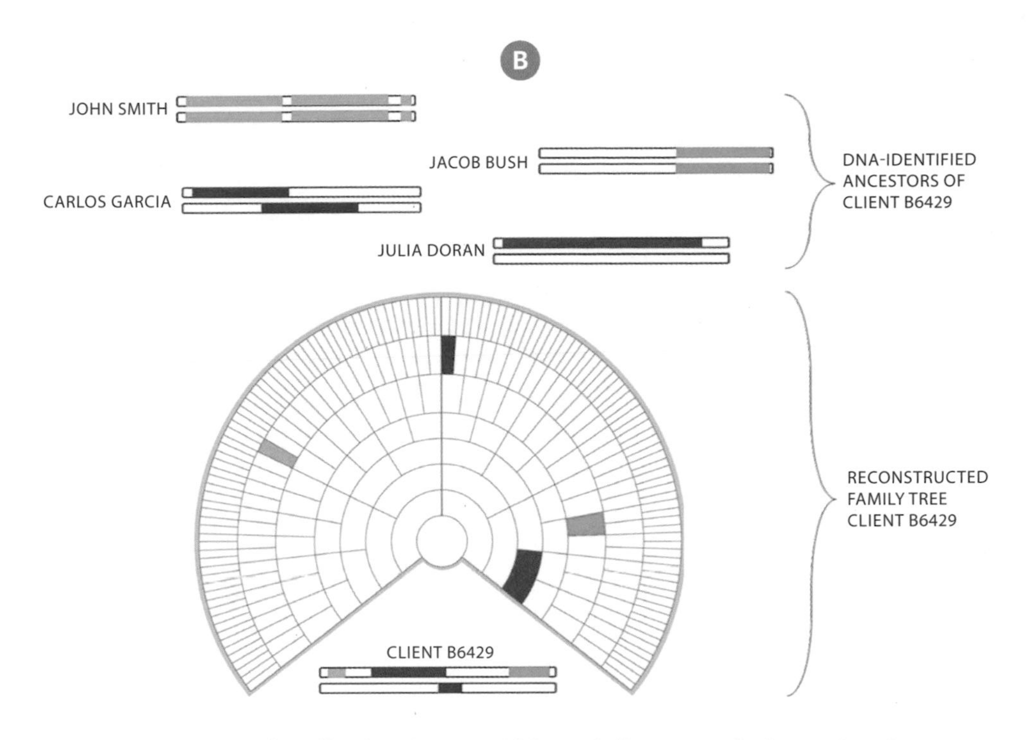

Future test-takers, like Client B6429 could theoretically construct family trees based on the segments of DNA they hold, inheritance patterns, and other factors.

single family tree; only a limited number of lines of descent lead from all of these twenty different ancestors to you. A future service could suggest having a certain relative tested (e.g., "We suggest having a descendant of your great-grandfather–your second cousin–tested to further refine your reconstructed family tree"), or ask you a series of questions to more accurately resolve the conflicts in the tree ("What was your maternal grandmother's name?" "What was your great-grandmother's name and date of birth?," and so on). By asking for user feedback, the program could select the most likely path of descent from your twenty ancestors to you and construct a likely family tree based on that information.

In the example shown in image B, Client B6429 possesses segments of DNA from four different reconstructed genomes. This information is used to create a reverse or reconstructed family tree with the identified ancestors mapped to it in the most likely configuration based on the size of the segments, established genealogies, and several other factors.

This reconstructed family tree process could also be used alongside traditional genealogical research, and in fact is used for identifying the families of adoptees. For example, a genealogist who knows a client is descended from ten people could easily recreate the probable family tree of that individual. Rather than creating a family tree entirely from scratch, the program pieces together portions of existing family trees in the database to generate possibilities for the client. While the most recent three to five generations might have to be filled in by the client (since these generations are the least likely to be included in the company's database), much of the tree could be completed based only on the DNA results.

There are, of course, many caveats with this method, and any computer-generated family tree should be confirmed by traditional research. Poor-quality trees, for example, will present a major challenge to this process, although they will not provide a complete stumbling block. Indeed, DNA evidence will likely ameliorate poor-quality trees to a significant extent. In addition to poor-quality trees, well-researched and well-documented family trees can be incorrect due to otherwise undetectable misattributed parentage events such as adoption, name change, or infidelity. These trees can also be detected and analyzed using the methods described above.

In addition, possessing DNA from a reconstructed genome does not automatically mean that the test-taker is descended from the person who possessed that genome. Instead, the test-taker may only be related to that person. For example, the test-taker may descend from the relatively unknown brother of John Smith who has very few living descendants, instead of being a descendant of John Smith himself. The test-taker could still possess any of the DNA hypothesized to be from John Smith. While the methods described above will likely ultimately focus on characterizing the genomes of "branch point ancestors" (such as immigrants, founders of unique haplotypes, and others) to avoid this problem, the existence of unknown or lesser known relatives of the branch point ancestor will temporarily throw a monkey wrench in the process.

Despite these caveats, ancestor and family tree reconstruction is likely to have an enormous impact on genealogical research in the next few decades, providing valuable ancestral information to users (especially adoptees).

Creating DNA-Only Ancestors

In the very near future, DNA from genetic cousins will be used to recreate the genomes of unknown ancestors who reside completely behind brick walls. While traditional research will often be able to provide a potential identity for the recreated genome, the individual will sometimes forever be known only by his reconstructed DNA. The DNA of these "DNA-Only Ancestors" is dispersed among living descendants, and some of it is already found within the testing companies' databases.

As we saw with the Speegles, it is possible to recreate at least a portion of an ancestor's genome if there are enough descendants. This process is greatly simplified if the family trees of those descendants are known and well researched. However, it will still be

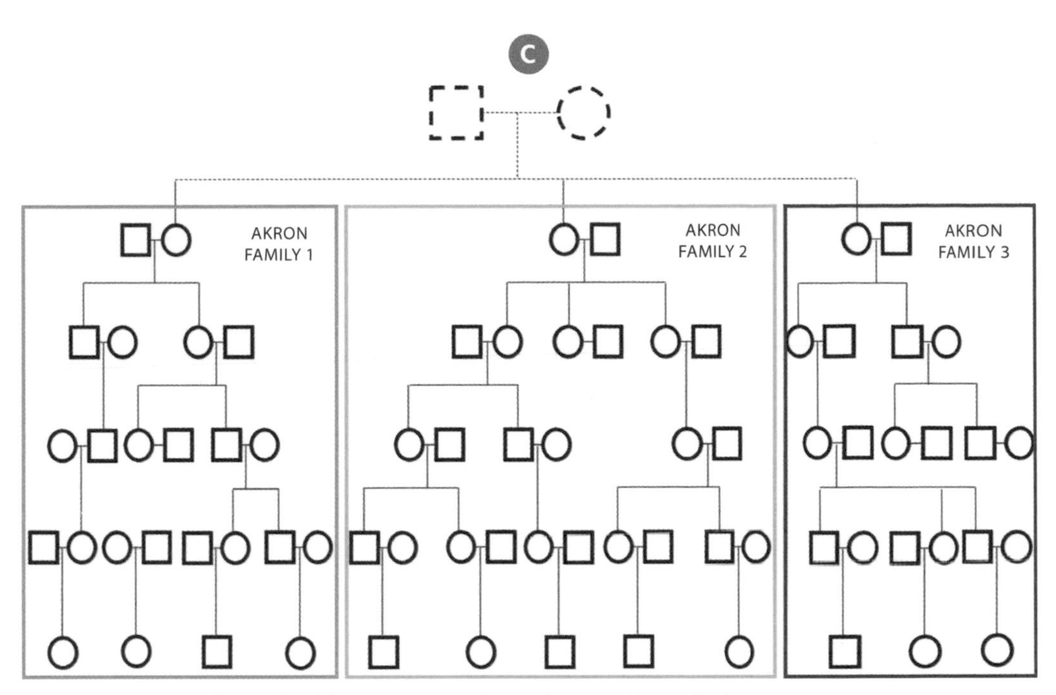

Shared DNA from suspected descendants could one day be combined
to piece together information about an ancestral couple.

possible to use the DNA of descendants of an ancestor who is not known to recreate portions of that unknown ancestor's genome.

Let's assume, for example, that a group of individuals trace their particular family tree to Akron, a small town in upstate New York in the early 1800s. The lines all end there with no known shared ancestor, and there are no traditional paper clues or shared surnames. However, all of these families are genetically related to one another, and based on their extensive research, they don't appear to share any other lines. If traditional research is exhausted, how can these relatives learn about their common ancestor?

Using the most advanced atDNA techniques, the DNA shared among the descendants could be assigned to an ancestor or ancestral couple (image C). The recreated partial genome will then provide other information about the DNA-only ancestor, such as predicted eye color, hair color, medical conditions, and traits. It could also be used to find other descendants or relatives.

Indeed, once a potential DNA-only ancestor is identified, some clues could help identify the name of the ancestor, such as inherent phenotypic information (a medical problem, for example) or other relatives who now show as a match because of the recreated genome (a Johnson with a strong oral record for a family in the same town, for example). Seeing a scholarly article in a national genealogy journal with the title "Pinpointing the

Likely Identity of a DNA-Reconstructed Ancestor in Akron, New York," is not as far away as you might think.

While it will be ideal to identify the name and family of the DNA-only ancestor, for many it will be impossible. This is especially true for regions and time periods where records are too sparse for such identification, such as eighteenth- and nineteenth-centuries in Ireland, African-American ancestry, Native American ancestry, and so on. For each of these regions, there will be many different DNA-only ancestors. While we may not know their names, we can fill in those gaps with any pieces of information we do have or complement the DNA-only identification with historical events or the life of a common individual in that time frame. For example, the profile for a DNA-only ancestor might look something like this:

- **AkronNY-1800s-Male-1:** Likely lived in Akron, South County, New York, between approximately 1800 and 1820. Had at least three children, probably daughters. Akron was first settled in 1797, so AkronNY-1800s-Male-1 was likely an early settler of the town, which prospered during its first two decades. Earliest known descendants are grandchildren Susannah (Unknown) Smith, Rebekah (Unknown) Mullen, and Sarah (Unknown) Johnson. AkronNY-1800s-Male-1 was of Irish descent and had blue eyes.

Although this example focuses on someone who was suspected of existing in a particular time and place, it will also be possible to recreate the genomes of individuals who were previously completely unknown to history and for which there are no paper or oral records of any kind.

Epigenetic Testing

All current DNA testing for genealogy looks at the sequence of *A* (adenines), *T* (thymine), *C* (cytosines), and *G*'s (guanines) along a chromosome or the mitochondrial genome. However, DNA comprises significant amounts of information beyond the order of *A*, *T*, *C*, and *G*'s. For example, in order to be a manageable size within the nucleus of the cell, DNA is packaged into a tight structure called **chromatin**, a complex bundle of DNA and packaging proteins. Some of the chromatin—called heterochromatin—is highly packaged and not actively used by the cell. Other portions of the chromatin—called euchromatin—is less tightly packaged and can be used by the cell. The packaging proteins themselves can be modified to affect how tightly packaged, and thus how active, portions of the genome are. In addition, the DNA itself can be tagged with chemical groups such as "methyl groups" that affect the accessibility and/or activity of that DNA (image D). Together, these epigenetic mechanisms have a direct and important impact on DNA activity.

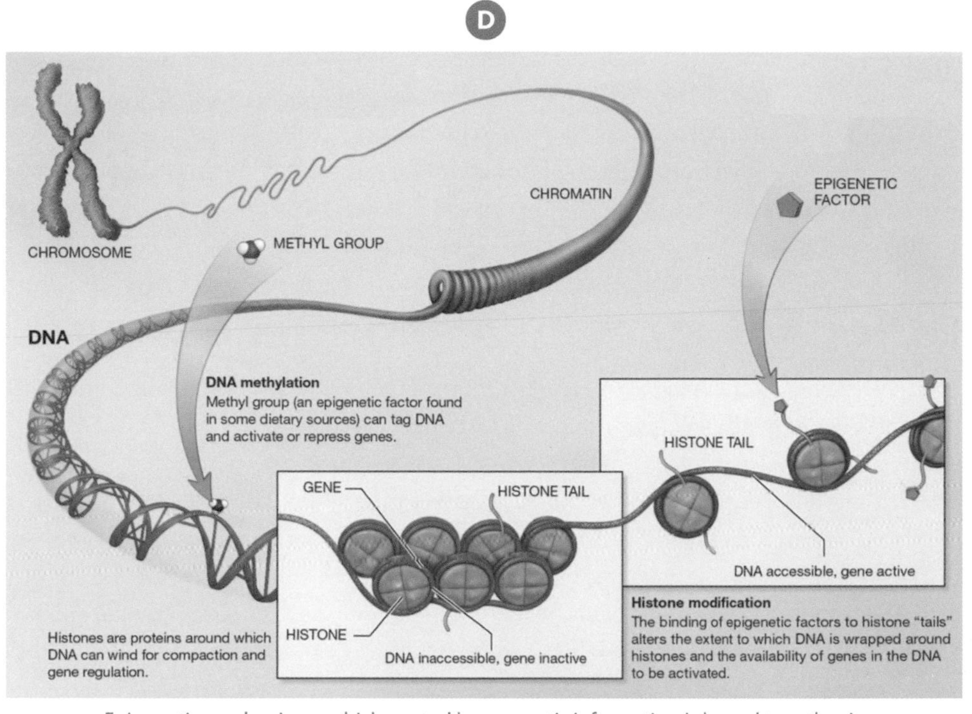

Epigenetic mechanisms, which control how genetic information is bound together in methyl groups that are arranged into chromatin, are beyond the scope of this book. However, future advancements in technology could uncover if and how this information can be useful to genealogists. Courtesy the National Institutes of Health.

Recent research has shown that at least some of the epigenetic structure of the DNA is likely inherited from one generation to the next <**www.discovermagazine.com/2013/may/13-grandmas-experiences-leave-epigenetic-mark-on-your-genes**>. For example, preliminary studies have suggested that people who experience trauma as a child—or had a parent or grandparent who experienced trauma—have different epigenetic profiles than those who did not experience similar trauma.

If epigenetic structure of DNA is indeed inherited, it can be utilized for genetic genealogical analysis. Although it is currently unclear whether epigenetic structure is stable and inherited for more than a handful of generations, it appears from current research that this epigenetic structure will at least be useful for examining recent and close relationships. For example, when determining whether someone is an aunt or half-sibling—or a third cousin versus a second cousin twice removed—epigenetic information might provide enough additional information to differentiate between possible relationships.

Additionally, epigenetic information might be useful to learn about the life experiences of our ancestors. There may be epigenetic markers of trauma (or of contentment) that have been inherited through generations. Once we assign a segment of DNA to an ancestor, for example, we might be able to characterize the epigenetics of that segment of DNA to learn about that ancestor or ancestral line.

Whether it is used only to differentiate between close cousin relationships or to learn about the life experiences of our ancestors, epigenetic analysis will almost certainly form an important part of genetic genealogy testing in the near future.

Other Advances in Genetic Genealogy

The developments discussed above are just some of the potential new tests or techniques that will benefit genealogy over the next few years. Of course, other improvements that will occur are impossible to predict.

Another, less predictable potential development in the field of genetic genealogy is improved cousin identification and characterization using both new testing technology such as whole-genome sequencing and massive family trees linked to DNA. For example, in the near future, cousins will likely be identified not only as a potential cousin, but as a statistically likely genealogical relationship, complete with an identified common ancestor. AncestryDNA's New Ancestor Discoveries attempts to do this, although it is a very early version.

Another area of development will be the increased use of DNA by lineage societies. As DNA becomes an increasingly important piece of genealogical evidence, it will continue to be adopted as evidence by lineage societies. The Daughters of the American Revolution, for example, announced a re-vamped DNA policy in 2013 that allows for limited uses of Y-DNA <see www.dar.org/national-society/genealogy/dna-and-dar-applications>. Other organizations are also allowing Y-DNA evidence to prove or support membership claims. In the future, however, results from other types of DNA tests might be available to create potential lineage lines for society members, particularly as genetic genealogists continue to demonstrate the power and efficacy of DNA testing for genealogy. Additionally, at some point, DNA evidence alone might be sufficient for membership, either because the DNA adequately establishes descent from an ancestor who qualifies for membership or because the lineage society itself is based on the DNA of particular ancestors.

CORE CONCEPTS: THE FUTURE OF GENETIC GENEALOGY

- Genetic genealogy is still a new field of scientific research. Future developments in both DNA testing technology and DNA analysis methodologies promise to add important new information to genealogical research.

- Improvements in Y-DNA sequencing will enable the discovery of new genealogically relevant Y-STRs and Y-SNPs to test and could enable even more refined estimates of paternal relationships.

- Affordable whole-genome sequencing of atDNA will allow for refined relationship predictions.

- Genealogists will recreate significant portions of their ancestors' genomes, revealing information about their lives, health, and physical appearance. Eventually, genealogists will be able to estimate the facial structure of ancestors.

- Some of these recreated genomes will belong to DNA-only ancestors who do not have a name or identity associated with them due to a lack of traditional genealogical records.

- Genealogists will be able to reconstruct portions of family trees from just the results of a DNA test, in conjunction with vast databases that combine family trees and DNA.

- Genealogists will use epigenetic testing to examine recent genealogical relationships and possibly to learn about the lives of our ancestors.

- Lineage societies will increase their acceptance of DNA evidence, and some may even rely entirely on DNA evidence.

Glossary

Genetic genealogy has a number of terms that may be unfamiliar to genealogists who haven't dabbled in DNA testing. To help you unpack these terms, this section contains a glossary of key terms used throughout the book, featuring a brief definition and a reference to the chapter in which the term first appears. Note these key terms may also appear in chapters other than the one referenced here.

admixture: Combination of different genetic lineages, usually with different geographic origins (chapter 9)

ancestral: Designation indicating the test-taker has an ancestral SNP value (i.e., no mutation) at a particular location (chapter 5)

autosomal DNA (atDNA): One of the four kinds of DNA useful to genealogists, found in the nucleus and comprising the twenty-two non-sex chromosomes (chapter 1)

autosome: One of the twenty-two non-sex chromosomes in the human genome (chapter 6)

Cambridge Reference Sequence (CRS): The first mtDNA sequence published; the long-time standard against which all test-takers' mtDNA are compared (chapter 4)

chromatin: Tight bundle of DNA and proteins that forms chromosomes; could be subject of future DNA testing (chapter 12)

chromosome: Structure containing millions of DNA base pairs; humans have a total of forty-six chromosomes, organized into twenty-three pairs (chapter 1)

chromosome browser: Tool that lets test-takers see exactly what segment(s) of their chromosomes are shared with another test-taker (chapter 6)

chromosome pair: Two complementary chromosomes that come together to form a pair (chapter 1)

coding region (CR): Region of mtDNA that contains genes and instructions for the cell and thus rarely changes; has only recently been included in mtDNA testing (chapter 4)

derived: Designation indicating the test-taker has a mutation at a particular SNP location (chapter 5)

DNA (deoxyribonucleic acid): Molecule that contains genetic information and can be a valuable tool for genealogists; two long chains containing millions of base pairs that form a double-helix structure (chapter 1)

ethnicity estimate: Method of inferring the geographical origins of an individual's DNA by comparing that DNA to one or more reference populations (chapter 9)

fully identical region (FIR): Portion of genome where two people share a segment of DNA on both of their chromosomes (chapter 6)

gene: Region of DNA that contains genetic information or instructions utilized by the cell, such as for the creation of proteins needed for life (chapter 1)

genealogical family tree: Collection of all an individual's ancestors, regardless of whether or not they contributed DNA to the individual (chapter 1)

genetic distance: Numerical representation of the differences or mutations between two individuals' Y-DNA or mtDNA results (chapter 4)

genetic exceptionalism: The theory that genetic information is unique and should be treated differently than other kinds of genealogical evidence (chapter 3)

genetic family tree: Collection of genealogical ancestors that contributed DNA to a genome; a subset of the genealogical family tree (chapter 1)

genetic genealogy: The practice and study of using DNA in genealogical research (chapter 1)

Genetic Genealogy Standards: Set of ethical principles and best-practices established by an ad hoc committee of scientists and genealogists (chapter 3)

half-identical region (HIR): Portion of genome where two people share a segment of DNA on just one of their two chromosomes (chapter 6)

haplogroup: Group of individuals who share several genetic mutations as well as a common (usually ancient) ancestor; exist on two genetic lines: mtDNA and Y-DNA (chapter 4)

haplotype: The collection of specific marker results that characterize a test-taker (chapter 5)

heteroplasmic: Containing more than one sequence of mtDNA within a cell or organism (chapter 4)

homoplasmic: Containing only one sequence of mtDNA within a cell or organism (chapter 4)

hypervariable control region 1 (HVR1): One of the two regions of mtDNA that frequently undergoes changes between generations and thus is often sampled in mtDNA testing (chapter 4)

hypervariable control region 2 (HVR2): One of the two regions of mtDNA that frequently undergoes changes between generations and thus is often sampled in mtDNA testing (chapter 4)

karyotype: All of the chromosome pairs in a human cell arranged in a numbered sequence from longest to shortest (chapter 1)

marker: An assigned, commonly tested region of DNA (chapter 2)

meiosis: A specialized process in which cells divide as eggs and sperm are created for reproduction (chapter 6)

mitochondria: Energy-producing units that live within cells and are inherited from the mother, where mtDNA is found (chapter 1)

mitochondrial DNA (mtDNA): One of the four kinds of DNA useful to genealogists, found in the mitochondria of a cell and always inherited from the mother (chapter 1)

most recent common ancestor (MRCA): The ancestor who is shared by two or more individuals and was born most recently (chapter 4)

mtDNA sequencing: One of the two kinds of mtDNA testing; examines part or all of an mtDNA nucleotide base pairs (chapter 4)

mutation: Any variance in DNA that occurs between individuals or between an individual and a reference sequence (chapter 4)

non-coding regions: Portions of the human genome that do not contain genetic information (chapter 1)

non-paternal events: Events or circumstances such as adoption, name change, or infidelity that lead to an unexpected break in a genetic line (chapter 2)

nonsister chromatids: Copies of non-identical chromosomes that have been duplicated during meiosis; any crossover/recombination between these two results in distinguishable mutations in the DNA (chapter 6)

nucleotide: Organic building blocks that form pairs to create DNA molecules; the four varieties are found in human DNA are adenine, cytosine, guanine, and thymine (chapter 1)

nucleus: Control center of cells, where most DNA is found (chapter 1)

phasing: Method of separating an individual's DNA into the DNA inherited from the mother and the DNA inherited from the father (chapter 9)

recombination: Process by which chromosome pairs exchange genetic material, leading to variations between generations (chapter 4)

Reconstructed Sapiens Reference Sequence (RSRS): Recent effort to represent a single mtDNA genome of all living humans; sometimes used as the standard against which test-takers' mtDNA is compared (chapter 4)

reference population: Group(s) of people to whom test-takers' results are compared (chapter 2)

revised Cambridge Reference Sequence (rCRS): An update to the CRS; commonly used as the standard against which test-takers' mtDNA is compared (chapter 4)

segment triangulation: Method of tracing one or more segments of an individual's DNA back to a specific ancestor or ancestral couple by comparing the DNA to that of two or more genetic relatives who all share the same segment of DNA and ancestor (chapter 10)

single nucleotide polymorphism (SNP): Single nucleotide in the DNA sequence that can differ between individuals in a population (chapter 4)

sister chromatids: Identical copies of a chromosome that has been duplicated during meiosis (chapter 6)

SNP testing: One of the two kinds of mtDNA testing; examines specific locations (SNPs) along the circular mtDNA molecule (chapter 4)

subclade: Subgroup of a haplogroup, defined by one or more SNP mutations (chapter 5)

tree triangulation: Method for finding ancestors by using other individuals' family trees in conjunction with genetic in-common-with tools (chapter 10)

whole-genome sequencing: Testing that examines all of an individual's DNA (chapter 6)

X-chromosomal DNA (X-DNA): One of the four kinds of DNA useful to genealogists, found on the X chromosome (chapter 1)

X chromosome: One of the two sex chromosomes that determine gender, among other traits; two X chromosomes (one inherited from each parent) result in the individual being female (chapter 7)

Y-chromosomal DNA (Y-DNA): One of the four kinds of DNA useful to genealogists, found on the Y chromosome that only males have and inherited only from the father (chapter 1)

Y-chromosome: One of the two sex chromosomes that determine gender, among other traits; one X chromosome (inherited from the mother) and one Y chromosome (inherited from the father) result in the individual being male (chapter 5)

Y-SNP testing: One of the two kinds of Y-DNA testing; examines specific locations (SNPs) along the Y-chromosome (chapter 5)

Y-STR testing: One of the two kinds of Y-DNA testing; examines short, repeating sequences of DNA (STR, or short tandem repeat) along the Y-chromosome (chapter 5)

A

Comparison Guides

There is no one-size-fits-all DNA testing plan, and for most of us the expense of DNA testing is a constant consideration. Although the cost of DNA testing has steadily declined, using multiple test types (or testing multiple people) still results in a considerable expense. If cost were not a factor, I would recommend a full mtDNA test, a 111-marker Y-DNA test (for males), and autosomal DNA tests at 23andMe **<www.23andme.com>**, AncestryDNA **<www.dna.ancestry.com>**, and Family Tree DNA **<www.familytreedna.com>**.

But because cost *is* a factor for most researchers, you'll need to be deliberate in selecting how you're going to allocate your resources. This section (containing a choosing-your-DNA-test flow chart, a table comparing the four major types of a testing, and a chart comparing the features of the three testing companies) is designed to help you select both a test and a testing company to accomplish your research goals and give you the largest bang for your buck.

The flowchart is a basic decision-making guide. It cannot cover all possible scenarios, and should not trump advice you've received from a company or someone with DNA testing experience. But if you have no idea where to start—and no one to ask—this flowchart will give you an idea of what test or tests you should pursue with just a few simple questions:

1. Are you testing to answer a specific genealogical question? In other words, are you testing to examine a specific relationship, brick wall, or mystery? If so, then you might require a more specific test such as a Y-DNA test (if it is a Y-DNA line) or an mtDNA test (if it is an mtDNA line). If not, and you're more interested in the general aspects of DNA testing, you should most likely start with an autosomal-DNA (atDNA) test. Not long ago, every genetic genealogist would have recommended starting with a Y-DNA or mtDNA test. Now, however, there is far more to work with once you receive your autosomal-DNA test results. Once you've explored your atDNA and want to experiment with more DNA testing, I recommend trying a Y-DNA test, then an mtDNA test.
2. Are you an adoptee? If yes, and you're a male adoptee, you should start with a Y-DNA test and also consider an atDNA test. If you're a female adoptee, head straight for an atDNA test.
3. Are you testing yourself? If yes, then you should consider an atDNA test at AncestryDNA and/or Family Tree DNA. If you're asking someone else to test, then proceed to the next question.
4. Is future testing a possibility? If you're asking another person to test and you'd like to have the sample available for future testing, especially if there is a possibility that the DNA provider won't be available, then consider testing with Family Tree DNA (or collecting a sample to be stored by Family Tree DNA).

In most cases requiring atDNA testing, I recommend that you test at AncestryDNA and Family Tree DNA. You can then consider testing at 23andMe, if the cost is not an issue. To decide for yourself, I've included a chart that compares the features of the three testing companies.

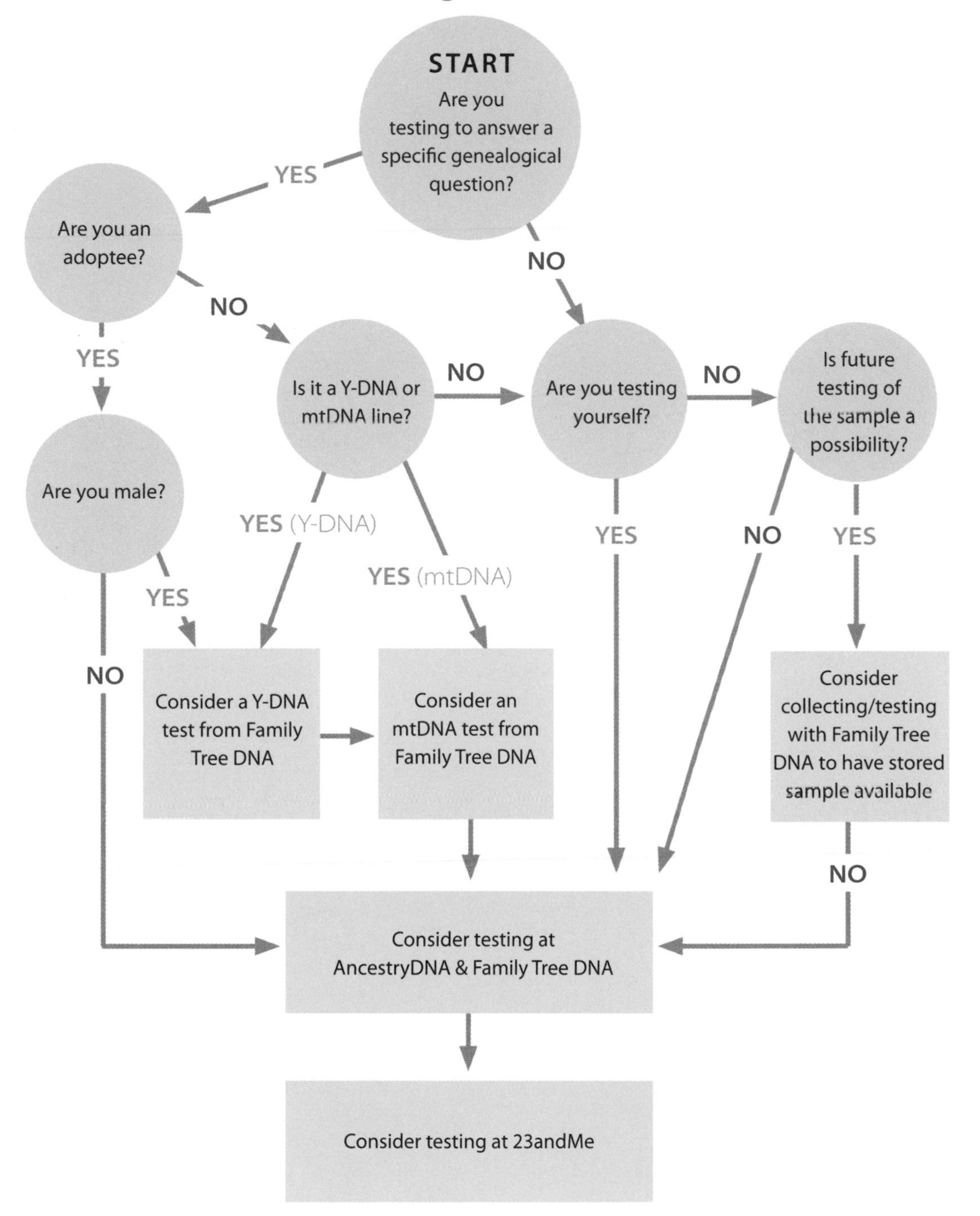
Choosing a DNA Test
START
Are you testing to answer a specific genealogical question?
YES
NO
Are you an adoptee?
NO
YES
Is it a Y-DNA or mtDNA line?
NO
Are you testing yourself?
NO
Is future testing of the sample a possibility?
Are you male?
YES (Y-DNA)
YES (mtDNA)
YES
NO
YES
YES
NO
Consider a Y-DNA test from Family Tree DNA
Consider an mtDNA test from Family Tree DNA
Consider collecting/testing with Family Tree DNA to have stored sample available
NO
Consider testing at AncestryDNA & Family Tree DNA
Consider testing at 23andMe

DNA Test Comparison

	mtDNA	Y-DNA	atDNA	X-DNA
Types of testing	• **HVR1/HVR2 sequencing:** Testing regions of DNA that are more likely to change • **Whole-mtDNA sequencing:** Testing the full mtDNA strand • **SNP testing:** Testing specific DNA sites	• **Y-STR testing:** Testing short repeated segments of DNA • **Y-SNP testing:** Testing specific DNA sites	• **SNP testing:** Testing specific DNA sites • **Whole-genome sequencing:** Testing all twenty-three chromosomes	(SNP testing is part of an atDNA test)
Haplogroup determination?	Yes	Yes. Y-DNA test results are used to either estimate (for Y-STR test) or determine (for Y-SNP testing) the test-taker's paternal haplogroup.	No	No
Cousin matching?	Yes. HVR1/HVR2 and whole-mtDNA sequencing can be used for cousin matching, although random matches may not be meaningful in a genealogically relevant timescale since mtDNA mutates slowly. SNP testing is not used for cousin matching.	Yes. Y-STR test results are useful for random cousin matching for estimating the number of paternal generations between two matches. Y-SNP testing is not as useful.	Yes. atDNA test results are useful for random cousin matching and for roughly estimating the number of generations between two matches.	Yes, although (due to low SNP density and low thresholds) only large segments should be considered (at least 10 cMs, and possibly larger)

Testing Company Comparison

		23andMe <www.23andme.com>	AncestryDNA <dna.ancestry.com>	Family Tree DNA <www.familytreedna.com>
General Information	price	$199	$99	$99 (atDNA); $169 (Y-DNA); $199 (mtDNA)
	database size	more than 1 million profiles	more than 2 million profiles	more than 700,000 profiles (atDNA, Y-DNA, and mtDNA combined)
	estimated message response rate of matches	low	medium	medium
	subscription required	no	yes, for some analysis tools	no
	accessibility to customer service	e-mail only	phone or e-mail	phone or e-mail
	contact your match	yes, directly	yes, via e-mail brokering	yes, directly
Genealogy Tools	search by surname	yes	yes	yes
	search by location	yes	yes	yes
	integrate pedigree with DNA	no	yes	no
Genetic Tools	possible relationship suggested	yes	yes	yes
	amount of shared DNA (in centimorgans)	yes	yes	yes
	chromosome browser	yes	no	yes
	see other matches shared with a match	yes	yes	yes

B

Research Forms

While it can be tempting to jump into research and cousin matching as soon as you've received your DNA results, you'll benefit from more careful and thorough analysis of your potential matches, web searches, and ancestral lines. This section contains a number of forms to help you analyze your research and keep your findings in order. You can download versions of these forms online at <**ftu.familytreemagazine.com/ft-guide-dna**>.

- **Relationship Chart:** Figure out how you're related to another person based on your most recent common ancestor.
- **Surname Worksheet:** Record important surname information for easy reference.
- **DNA Cousin Match Worksheet:** Document your confirmed DNA cousins.
- **Match Relationships Worksheet:** Determine how you and potential matches relate.
- **Family Group Sheet:** List all you know (and discover) about a particular family.
- **Five-Generation Ancestor Chart:** Trace your family tree back five generations.
- **Research Log and Planner:** Track what you've accomplished in your research—and what you still need to do.

Relationship Chart

Instructions:

1. Identify the most recent common ancestor of the two individuals with the unknown relationship.
2. Determine the common ancestor's relationship to each person (for example, grandparent or great-grandparent).
3. In the topmost row of the chart, find the common ancestor's relationship to cousin number one. In the far-left column, find the common ancestor's relationship to cousin number two.
4. Trace the row and column from step 3. The square where they meet shows the two individuals' relationship.

THE MOST RECENT COMMON ANCESTOR IS COUSIN NUMBER ONE'S ...

THE MOST RECENT COMMON ANCESTOR IS COUSIN NUMBER TWO'S ...	parent	grandparent	great-grandparent	great-great-grandparent	third-great-grandparent	fourth-great-grandparent	fifth-great-grandparent	sixth-great-grandparent
parent	siblings	nephew or niece	grandnephew or -niece	great-grandnephew or -niece	great-great-grandnephew or -niece	third-great-grandnephew or -niece	fourth-great-grandnephew or -niece	fifth-great-grandnephew or -niece
grandparent	nephew or niece	first cousins	first cousins once removed	first cousins twice removed	first cousins three times removed	first cousins four times removed	first cousins five times removed	first cousins six times removed
great-grandparent	grandnephew or -niece	first cousins once removed	second cousins	second cousins once removed	second cousins twice removed	second cousins three times removed	second cousins four times removed	second cousins five times removed
great-great-grandparent	great-grandnephew or -niece	first cousins twice removed	second cousins once removed	third cousins	third cousins once removed	third cousins twice removed	third cousins three times removed	third cousins four times removed
third-great-grandparent	great-great-grandnephew or -niece	first cousins three times removed	second cousins twice removed	third cousins once removed	fourth cousins	fourth cousins once removed	fourth cousins twice removed	fourth cousins three times removed
fourth-great-grandparent	third-great-grandnephew or -niece	first cousins four times removed	second cousins three times removed	third cousins twice removed	fourth cousins once removed	fifth cousins	fifth cousins once removed	fifth cousins twice removed
fifth-great-grandparent	fourth-great-grandnephew or -niece	first cousins five times removed	second cousins four times removed	third cousins three times removed	fourth cousins twice removed	fifth cousins once removed	sixth cousins	sixth cousins once removed

Surname Worksheet

Surname		Soundex Code
Meaning		
Spelling Variations		
Possible Transcription Errors		

Surname		Soundex Code
Meaning		
Spelling Variations		
Possible Transcription Errors		

Surname		Soundex Code
Meaning		
Spelling Variations		
Possible Transcription Errors		

DNA Cousin Match Worksheet

Percentage Match	Centimorgans (CM)	Relationship	Notes

Match Relationships Worksheet

Page 1

Use this tracker to note key clues that could help you determine how you and your genetic cousins are related.

	Testing Company and Website	Username of Match	Estimated Relationship	Contact Info (If Known)	Shared Ancestral Places	Match's Ancestors from Shared Places
1						
2						
3						
4						
5						
6						
7						

Match Relationships Worksheet

	Shared Surnames	Match's Relative(s) with That Surname (and Relationship to User)	Shared Ethnic Origins	Correspondence with User, Including Dates	Notes
1					
2					
3					
4					
5					
6					
7					

Family Group Sheet

of the________________________________Family

	Source #		Source #
Full Name of Husband		Birth Date and Place	
His Father		Marriage Date and Place	
His Mother with Maiden Name		Death Date and Place — Burial	
Full Name of Wife			
Her Father		Birth Date and Place	
Her Mother with Maiden Name		Death Date and Place — Burial	
Other Spouses		Marriage Date and Place	

Children of This Marriage	Birth Date and Place	Death Date, Place and Burial	Marriage Date, Place and Spouse

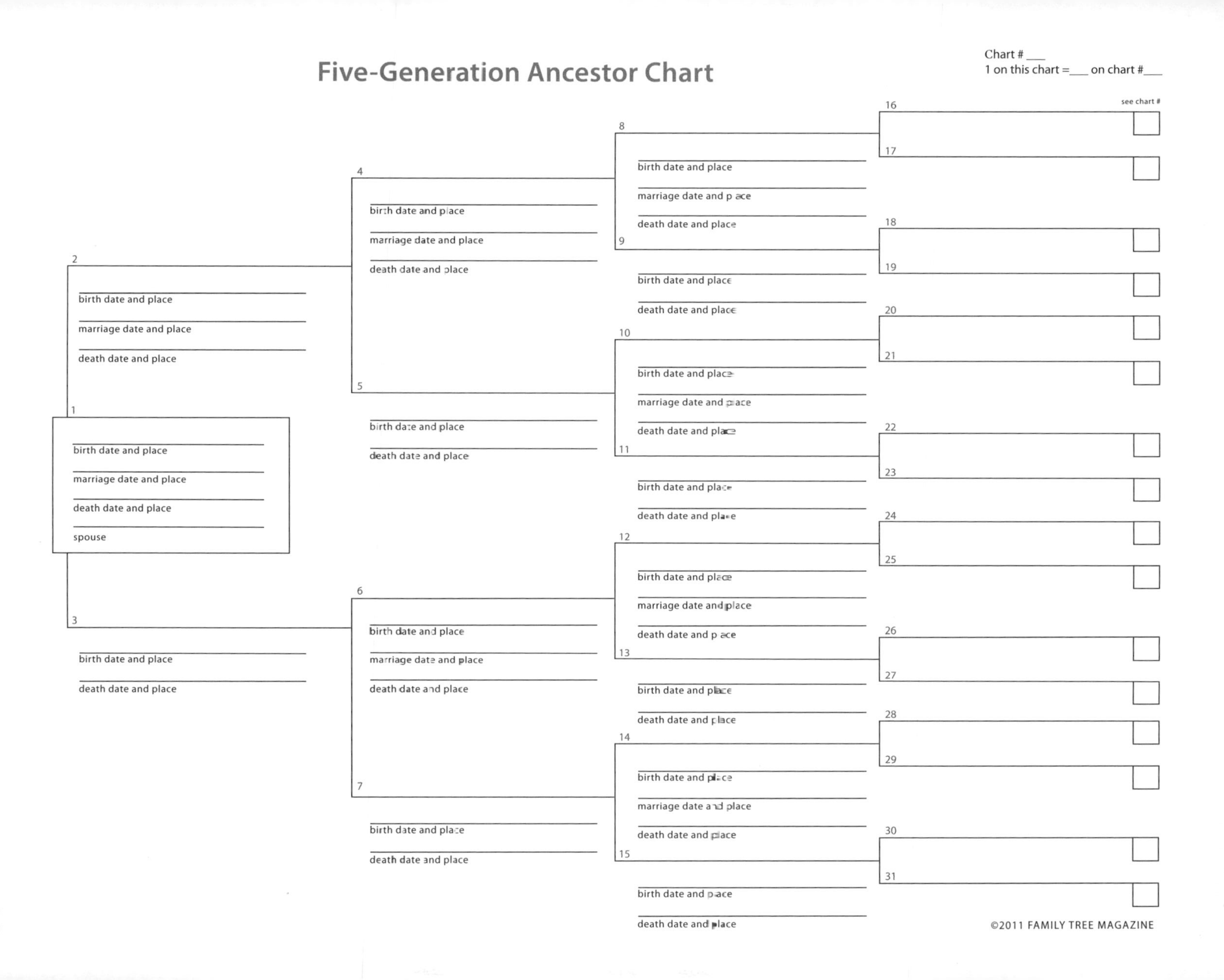
Five-Generation Ancestor Chart
Chart # ___
1 on this chart = ___ on chart # ___
see chart #
1
birth date and place
marriage date and place
death date and place
spouse
2
birth date and place
marriage date and place
death date and place
3
birth date and place
death date and place
4
birth date and place
marriage date and place
death date and place
5
birth date and place
death date and place
6
birth date and place
marriage date and place
death date and place
7
birth date and place
death date and place
8
birth date and place
marriage date and place
death date and place
9
birth date and place
death date and place
10
birth date and place
marriage date and place
death date and place
11
birth date and place
death date and place
12
birth date and place
marriage date and place
death date and place
13
birth date and place
death date and place
14
birth date and place
marriage date and place
death date and place
15
birth date and place
death date and place
16
17
18
19
20
21
22
23
24
25
26
27
28
29
30
31
©2011 FAMILY TREE MAGAZINE

Research Planner and Log

Research question:

____

____

Known Information:

____

____

TASK	DONE?	RESULT/COMMENTS	EXPENSES

C

More Resources

A genealogist's education is never complete. If you've mastered the subject matter in this book, I recommend that you find other resources to explore. The best way to learn about genetic genealogy is to test yourself and family members, and work with the results as much as possible. In addition to testing, here are a few of the best resources available to genealogists interested in learning more about DNA.

These are just a few of the blogs, websites, forums, and mailing lists dedicated to genetic genealogy. In addition to these resources, DNA is now an essential topic at every genealogy conference in the United States, be sure to attend local conferences as well.

ISOGG Wiki

The International Society of Genetic Genealogy (ISOGG) Wiki **<www.isogg.org/wiki/Wiki_Welcome_Page>** is an essential resource for genetic genealogists. Although it is a Wikipedia-style source of information curated by volunteers, it contains some of the most sophisticated and detailed analysis of topics related to genetic genealogy. The following pages, for example, are required reading for genetic genealogists:

- Autosomal-DNA statistics <**www.isogg.org/wiki/Autosomal_DNA_statistics**>
- Autosomal-DNA testing comparison chart <**www.isogg.org/wiki/Autosomal_DNA_testing_comparison_chart**>
- Ethics, guidelines and standards <**isogg.org/wiki/Ethics,_guidelines_and_standards**>
- Identical by descent <**www.isogg.org/wiki/Identical_by_descent**>
- Triangulation <**isogg.org/wiki/Triangulation**>

Books

In addition to the book you're holding in your hand (or reading on your screen!), there are several other books dedicated to the fundamentals of genetic genealogy.

- Blaine Bettinger and Debbie Parker Wayne, *Genetic Genealogy in Practice* (Arlington, Va.: National Genealogical Society, 2016).
- David R. Dowell, *NextGen Genealogy: The DNA Connection* (Libraries Unlimited, 2014).
- Debbie Kennett, *DNA and Social Networking: A Guide to Genealogy in the Twenty-first Century* (Gloucestershire, United Kingdom: The History Press, 2011).
- Emily D. Aulicino, *Genetic Genealogy: The Basics and Beyond* (Bloomington, Ind.: AuthorHouse, 2013).
- Richard Hill, *Guide to DNA Testing: How to Identify Ancestors and Confirm Relationships through DNA Testing* (2009). <**www.dna-testing-adviser.com/DNA-Testing-Guide.html**>

Blogs

Blogs are a great way to stay on top of the latest developments in the field. Here is an essential list of the best blogs for genetic genealogists. Although many of these blogs are not updated frequently, they all contain archives full of very rich content and information.

- *23andMe Blog* <**blog.23andme.com**>
- AncestryDNA Blog <**blogs.ancestry.com/ancestry/category/dna**>
- *Cruwys News* <**cruwys.blogspot.com**> by Debbie Kennett
- *Deb's Delvings in Genealogy* <**debsdelvings.blogspot.com**> by Debbie Parker Wayne
- *DNAeXplained—Genetic Genealogy* <**dna-explained.com**> by Roberta Estes
- *Dr D Digs Up Ancestors* <**blog.ddowell.com**> by David R. Dowell

- *Genealem's Genetic Genealogy* <genealem-geneticgenealogy.blogspot.com> by Emily Aulicino
- *Genealogy Junkie* <www.genealogyjunkie.net/blog> by Sue Griffith
- *The Genetic Genealogist* <www.thegeneticgenealogist.com> by Blaine Bettinger
- *Kitty Cooper's Blog: Musings on Genealogy, Genetics, and Gardening* <blog.kitty-cooper.com> by Kitty Cooper
- *The Lineal Arboretum* <linealarboretum.blogspot.com> by Jim Owston
- *Segment-ology* <segmentology.org> by Jim Bartlett
- *Through the Trees* <throughthetreesblog.tumblr.com> by Shannon Christmas
- *Your DNA Guide* <www.yourdnaguide.com> by Diahan Southard
- *Your Genetic Genealogist* <www.yourgeneticgenealogist.com> by CeCe Moore

Forums and Mailing Lists

Forums and mailing lists encourage interaction, questions, and conversation. Many of these forums and mailing lists can be set up such that you can monitor them without receiving numerous daily emails.

- 23andMe Forums (23andMe) <www.23andmeforums.com>
- Anthrogenica Forums (Anthrogenica) <www.anthrogenica.com/forum.php>
- DNAAdoption (Yahoo! Groups) <groups.yahoo.com/neo/groups/DNAAdoption/info>
- DNA Detectives (Facebook) <www.facebook.com/groups/DNADetectives>
- DNAgedcom User Group (Facebook) <www.facebook.com/groups/DNAGedcomUserGroup>
- DNA Newbie (Facebook) <www.facebook.com/groups/dnanewbie>
- DNA: GENEALOGY—DNA mailing list (Rootsweb) <lists.rootsweb.ancestry.com/index/other/DNA/GENEALOGY-DNA.html>
- DNA-NEWBIE (Yahoo! Groups) <groups.yahoo.com/neo/groups/DNA-NEWBIE/info>
- Family Tree DNA Forums (Family Tree DNA) <forums.familytreedna.com>
- International Society of Genetic Genealogy—ISOGG (Facebook) <www.facebook.com/groups/isogg>
- GEDmatch User Group (Facebook) <www.facebook.com/groups/gedmatchuser>

INDEX

ABOUT THE AUTHOR

Blaine Bettinger Ph.D. (biochemistry), J.D. is an intellectual property attorney at Bond, Schoeneck & King, PLLC in Syracuse, New York, by day, and a genealogy educator and blogger by night. In 2007 he created *The Genetic Genealogist* <**www.thegeneticgenealogist.com**>, one of the first blogs devoted to genetic genealogy and personal genomics.

Blaine has written numerous DNA-related articles for the *Association of Professional Genealogists Quarterly, Family Tree Magazine*, and other publications. He has been an instructor at the inaugural genetic genealogy courses at the Institute of Genealogy and Historical Research (IGHR), Salt Lake Institute of Genealogy (SLIG), Genealogical Research Institute of Pittsburgh (GRIP), Virtual Institute of Genealogical Research, Family Tree University, and Excelsior College (Albany, NY). He is a former editor of the *Journal of Genetic Genealogy*, and a co-coordinator of the ad hoc Genetic Genealogy Standards Committee. In 2015, he became an alumnus of ProGen Study Group 21 and was elected to the New York Genealogical and Biographical Society's Board of Trustees.

Blaine was born and raised in Ellisburg, NY, where his ancestors have lived for more than two hundred years, and is the father of two boys. You can find Blaine at <**www.blainebettinger.com**> and on Twitter *@blaine_5*.

DEDICATION

To my paternal grandparents Roy Harry and Laurentine Loverna (Mullin) Bettinger, and my maternal grandparents Theodore Roosevelt and Jane Rose (Garcia) LaBounty.

PHOTO CREDITS

Acknowledgments

When I took my first DNA test in 2003—or when I started blogging about DNA in 2007—I had no idea that it would lead to so many incredible opportunities, including this book. My rewarding relationship with *Family Tree Magazine* began in 2009, and I am deeply indebted to the entire team at F+W, past and present, including Diane Haddad, Tyler Moss, Allison Dolan, Andrew Koch, Vanessa Wieland, and everyone else. The guidance, advice, and encouragement I received from F+W throughout this process made everything possible. Thank you.

Thank you to 23andMe **<www.23andme.com>**, AncestryDNA **<www.dna.ancestry.com>**, Family Tree DNA **<www.familytreedna.com>**, GEDmatch **<www.gedmatch.com>**, and DNAGedcom **<www.dnagedcom.com>** for everything you do for the community, and for allowing me to use screenshots for the book.

Thank you to my wonderful friends and colleagues in the genealogical community, who encourage me and inspire me on a daily basis. In particular, thank you to my fellow institute instructors CeCe Moore, Debbie Parker Wayne, and Angie Bush, from whom I've learned so much. And my sincere appreciation to the many genetic genealogists and educators all around the world who collaborate, share, and debate issues great and small in order to move our understanding of genetic genealogy forward every day.

Thank you to my middle school English teacher, Mrs. Briant, who unknowingly but irreversibly changed my world with a simple assignment to fill out a family tree. It's an assignment I'll be working on for the rest of my life.

Thank you to my parents, brothers, and sister who have supported and encouraged my genealogy addiction for decades now, and have even spit for a DNA test or two. The biggest thank you, of course, goes to Elijah and Logan. They sacrifice so much so that I can travel, teach, lecture, and write. Someday I hope they look back and remember not a missed evening here or there, but that I was doing what I loved and that they should do the same.

 Published by Family Tree Books, an imprint of F+W Media, Inc., 10151 Carver Road, Suite 200, Blue Ash, Ohio 45242. (800) 289-0963. First edition.

ISBN: 978-1-4403-4532-6

20 19 18 17 16 5 4 3 2 1

DISTRIBUTED IN CANADA BY FRASER DIRECT
100 Armstrong Avenue
Georgetown, Ontario, Canada L7G 5S4
Tel: (905) 877-4411

DISTRIBUTED IN THE U.K. AND EUROPE BY
F&W Media International, LTD
Brunel House, Forde Close,
Newton Abbot, TQ12 4PU, UK
Tel: (+44) 1626 323200,
Fax (+44) 1626 323319
E-mail: enquiries@fwmedia.com

fw a content + ecommerce company

PUBLISHER AND COMMUNITY LEADER: Allison Dolan

EDITOR: Andrew Koch

DESIGNER: Julie Barnett

PRODUCTION COORDINATOR: Debbie Thomas

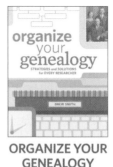